DICTIONARY OF
SOIL SCIENCE

DICTIONARY OF
SOIL SCIENCE

by
Dr. R.P. Chandola

Daya Publishing House®
A Division of
Astral International Pvt. Ltd.
New Delhi – 110 002

First Published, 2010

ISBN 978-81-7035-655-4

Publisher's Note:

Every possible effort has been made to ensure that the information contained in this book is accurate at the time of going to press, and the publisher and author cannot accept responsibility for any errors or omissions, however caused. No responsibility for loss or damage occasioned to any person acting, or refraining from action, as a result of the material in this publication can be accepted by the editor, the publisher or the author. The Publisher is not associated with any product or vendor mentioned in the book. The contents of this work are intended to further general scientific research, understanding and discussion only. Readers should consult with a specialist where appropriate.

Every effort has been made to trace the owners of copyright material used in this book, if any. The author and the publisher will be grateful for any omission brought to their notice for acknowledgement in the future editions of the book.

Published by : **Daya Publishing House®**
A Division of
Astral International Pvt. Ltd.
– ISO 9001:2015 Certified Company –
4736/23, Ansari Road, Darya Ganj
New Delhi-110 002
Ph. 011-43549197, 23278134
e-mail: info@astralint.com
Website: www.astralint.com

Digitally Printed at : **Replika Press Pvt. Ltd.**

Preface

Words in any language have always taken a major role in the advancement of human civilization. With the passing of time many words automatically added and enriched the various cultures towards provisions of a better, secured and longer life of all the living beings on this planet–the earth. Words have been used in the area of education which in turn has played a significant role towards improvement and upgradation of human life. In this regard scientific knowledge acquired through modern scientific research and education has influenced the human life tremendously vis-à-vis all other connected human activities.

Agricultural science including animal science as a particular way of life as adopted by human beings and also moved in the same direction as other sciences and also acts made good progress only primarily by means up many words with passing out of time. As far as scientific development for further improvement is concerned, it firstly evolved the scheme of making dictionaries to make it possible for easy comprehension of human behaviour, actions and reactions towards a civilized society. Such works also helped and emerged as one of the most important way in achieving an easy consultation for an achievement of rapid and significant knowledge of the different disciplines of agricultural science. In the present vastness of ocean of knowledge of any subject, it is difficult and cumbersome find the correct words for the different functions of agricultural science. The situation has become more tedious as the worlds pertain to the scientific and technical notations of various agricultural branches of knowledge have become numerous.

With the allure in view, six volumes relevant to the various subjects of agriculture *e.g.* Animal Science, Entomology (insect pest study), Crop Science (growing various agricultural field crops, agronomy), Plant Science (plant breeding and genetics), Plant Pathology (plant diseases), and Soil Science have been prepared with all scientific names (terms) used in agricultural education and research with comprehensive notes for various terms. Whenever possible some entries have been illustrated to make an easy understanding possible. Such a dictionary should also be used as a quick and easy reference to students and other concerned groups.

Dr. R.P. Chandola

A

Ablation The process in which snow and ice is lost from the surface of a nice sheet or glacier. This may be through melting, sublimation (the direct transference of water to the atmosphere from the solid to a gaseous) or abrasion.

Abney level An instrument used in surveying to measure the degree of slope from an observer to an object. It consists of a small telescope linked to a semicircular protractor, and a pointer with a spirit level attached. An object sighted through the telescope that is not on a level with the observer will throw the bubble in the spirit level out of centre. By adjusting a screw the bubble can be brought into view in the telescope. The pointer then shows how much the telescope is tilted from the horizontal plane.

Abrasion The wearing down of a land surface by moving weathered rock debris. This debris may be transported by water (*e.g.* by rivers, ice, or sea) or by wind.

Abrasion platform A level rock platform cut into a slope or cliff. It is most commonly a feature of coastal areas where it is known as a wave-cut platform and is generally produced in hard resistant rocks.

Absolute humidity *See* crop science.

Absorption A process in which a gas is taken up by a liquid or solid or in which a liquid is taken up by a solid in absorption, the substance absorbed goes into the bulk of the material. Solids that absorb gases or liquids often have a porous structure. The absorption of gases in solids is sometimes called *sorption.*

Absorption indicator An indicator used for titrations that involve a precipitation reaction. The method depends upon the fact that at the equivalence point there is a change in the nature of the ions absorbed by the precipitate particles. Fluorescein-a fluorescent compound–is generally used.

Accelerated soil erosion One of the outstanding causes of accelerated soil erosion due to water commonly seen everywhere, is the unwise deforestation and the subsequent neglect of such lands. A well-established forest not only intercepts the rainwater by its foliage but also develops a soil cover of organic matter. This humus layer, often quite deep, has a great absorptive capacity and tends to retard the rate of runoff as well as to soil in place and thus prevent erosion even when the runoff is great. Since much of the forestland is steep, the removal of the forest cover usually results in very rapid erosion. Even on fairly smooth upland, the surface layer of soil may be removed before the stumps are rotted away.

Acceleration It is the rate of change of velocity. It is calculated by subtracting the initial velocity from the final velocity and dividing it by time (rectilinear motion).

Acceleration due to gravity–symbol $g = \frac{GM}{R^2}$. The value of g is maximum at the poles and g is maximum at the poles and minimum at the equator.

Acclimatize *See* crop science.

Accordant drainage A pattern of drainage that closely reflects the rock type and structure over which the streams flow.

Accumulated temperature A measure of the variation of temperature from a given norm for a particular place over a specified period, measured in degree days. This is most common used for determining whether an area is suitable for the cultivation of particular crops. Over a given period, say 30 days, the difference between each mean daily temperature below which plant growth is inhibited, are added together, the result indicating the indicating effectiveness of temperature over the given period.

Acid A substance that gives rise to hydrogen ions (or H_3O^+) when dissolved in water. An acid in solution will have a pH below 7. This definition does not take into account the competitive behaviour of acid in solvents and it refers only to aqueous systems. Strong acids (*e.g.* NHO_3^+) react completely with water to give HNO_3^+), *i.e.* HNO_3 is stronger than H_3O^+) and the conjugate base NO_3 is weak. Weak acids (*e.g.* CH_3COOH and C_6H_5COOH) are only partly dissociated because H_3O^+ is a stronger acid than the free acids and the ions CH_3COO^- and $C_6H_5COO^-$ are moderately strong bases.

Acid anhydride An organic compound of general formula RCCO-COR′ where R and R′ are alkyl or aryl groups. These are prepared by reaction of an acyl halide with the sodium salt of a carboxylic acid *e.g.* $RCOCl + R'COO^-Na^+ \rightarrow RCCOO\text{-}COR'^+ NaCl$. Like the acyl halides, they are very reactive acylating agents. Hydrolysis is to carboxylic acids: $RCOOCOR'^+ H_2O \rightarrow RCOOH + R'COOH$, $RCOOH + SOCl_2 \rightarrow RCOCl + SO_2 + HCl$ acids halides have irritating vapours and fume in moist air. They are very reactive to the hydrogen atom of compounds containing hydroxyl (–OH) or amine ($-NH_2$) groups.

Acidic hydrogen *See* acid.

Acidimetry Volumetric analysis or acid-base titration in which a standard solution of an acid is added to the unknown (base) solution plus the indicator. Alkalimetry is the converse, *i.e.* the base is in the burette.

Acid lava a stiff viscous molten rock that flows slowly, has a high melting point (about 850°), and congeals into thick tongues before having travelled far from the volcanic vent. The resulting landform is a steep-sided one, typically represented by the Puys of the Auvergne, France. Acid lava is rich in silica (at least 66 per cent), and contains quartz crystals (at least 10 per cent). Examples from the geological record can be found in Cumbria, Fife, and snowdonia.

Acid rock An igneous rock that is very rich in silica (over 66 per cent) and aluminium, and which has at least 10 per cent of its composition made up of quartz crystals. It is relatively light and forms the main material of the continents. Examples of acid rocks include granite, obsidian, and rhuolite.

Acid salt (acidic salt) A salt is which there is only partial replacement of the acidic hydrogen of an acid by metal or other cations. For polybasic acids the formulae are of the type $NaHSO_4$ (sodium hydrogensulphate) and $Na_3H(CO_3)_2.2H_2O$ (sodium sesquicarbonate). For monobasic acides such as HF the acid salts are of the form KHF_2 (potassium hydrogen fluoride). Although the latter where at one time formulated as a normal salt plus excess acid (*i.e.* KF.HF, it is preferable to treat these a hydrogen-bonded systems of the type $K^+(F\text{-}H\text{-}F)7$.

Acid soils In humid regions, where rainfall is high, the soluble bases formed in the course of weathering of various siliceous rocks are carried away by drainage waters. Continued leaching results in the washing away of calcium, magnesium, potassium and sodium ions and the accumulation of relatively insoluble acid residues, composed largely of silicon, aluminium and iron in the form of oxides or silicate combinations of oxides. Such soils are acidic in reaction and they are called acid soils. Continual removal of lime and other base elements by crops and the accumulation of acids contained in certain manures (particularly sulphate of ammonia) also tend to produce acidic conditions. Besides mineral exchange materials, soils also contain organic exchange materials which when saturated with hydrogen, constitute the so-called humic acids. Acidity is also caused if soils develop on acid rocks.

Actinomyces These organisms are undoubtedly of great importance in respect to the dissolution of soil organic matter and the liberation of nutrients therefrom. Apparently, they reduce, rapidly and successfully, all sorts of compounds to simpler forms, even the lignins, the most resistant of all the organic compounds. Their optimum development in the soil lies between 6.0 and 7.5 pH and is practically inhibited at a pH 5.0 or below.

They are similar in size to bacteria but resemble molds in their grown and physiology. They can crow in deeper layers of soil and under drier conditions, and need less nitrogen. They are also said to synthesize chemical substance which produces the familiar odour, when the sunbaked soil of a cultivated field receives the first shower of rain.

Active fault A break in the earth's crust along which movement is taking place. Earthquakes occur as a result of this movement.

Active layer In permafrost regions, the top layer of soil that seasonally thaws; this can vary in depth from a few centimetres to several metres. The active layer is important for the production of a wide range of periglacial landforms.

Active lime In humid region mineral soils the amount of calcium in an exchangeable form is usually six or eight times greater than the potassium and three or four times greater than the magnesium. Moreover, as much as 40 per cent of total calcium may be exchangeable while 2 per cent ordinarily is a high figure for potash and 1.0 per cent for magnesium. As commonly known, lime stimulates nitrification in soils, even in those that may already contain a fair amount of active calcium.

Active portion of soil All normal soils are dynamic and the markedly active constituents either are in a finely dispersed condition or are in contact with materials already in this state. In other words, most soil reactions, physical, chemical and biological depend in large degree on the colloidicity of certain major constituents, both living and non-living.

Active volcano A volcano that continually or periodically erupts. Compare dormant volcano, extinct volcano.

Activity Symbol A: The average number of atoms disintegrating per unit time in a radioactive substance.

Adiabatic process An atmospheric process in which changes take place in the pressure and volume of a body of gas without an actual loss or gain of heat from outside. As a consequence the temperature of the air mass changes.

Adobe It is a term applied to a fine calcareous clay or silt and is very productive where irrigation is available.

Adret A slope that faces towards equator, therefore receiving more sunshine than slopes that face away from the Sun. Adret slopes tend to be better drained and warmer and have an important influence on agriculture, especially in mountainous regions.

Adsorption The adhesion of particles of one substance to the surface of another, such as the attraction of liquids or gases to a solid surface. In soils, this includes the concentration of base salts (*e.g.* calcium and potassium) of microscopic size (colloids) surrounding larger mineral particles. Adsorption larger mineral particles. Adsorption also refers to the process by which a thin film of water surrounds individual articles in sediments.

Adsorption by soils An ionic adsorption and exchange takes place and in some cases assumes considerable importance. This is true especially in respect to the adsorption of phosphate ions.

Advection The horizontal movement of an air mass across the earth's surface. Such movement is especially important when the air mass is moving from a different source are and is bringing in different conditions of temperature and humidity. Coastal and frontal areas often experience this and advection fog frequently resulted.

Aeolian Denoting the processes of landforms caused by or related to the wind. These processes or forms are most significant in arid and semiarid regions, but can also be important in some coastal and periglacial environments. Also wind deposits; easy to keep in good tilth; as coastal sand dunes, of not much agricultural value; found in parts of desert and south-west of India.

Aeolian deposits These comprise of all such material such as loses, abode and volcanic ash and sand dunes.

Aeration An important aspect of soil water. Since the gravitational water occupies the large pore spaces of the soil, it excludes air and interferes with oxidation. A poorly drained soil becomes soon unsatisfactory for the production of most crops. The normal soil processes, both chemical and biological are slowed down and favourable physical relations are weakened. In fact, in the absence of proper soil aeration, the whole cycle of events on which most higher plaints depends is thrown out of adjustment.

Age The smallest division of geological time. The rocks laid down during an age are known as a formation. The number of ages together comprise an epoch.

Age of soil As the influence becomes constructive in a soil sense and the presence of organic matter begins to exert its influence, a young soil gradually develops acquired characteristics, *i.e.* those included largely by climate, are noticeable but those inherited from the parent material are still in the ascendancy. The profile, however, since the constructive agencies have not had time to impress their patterns, is by no means fully developed.

Agglomerate A rock made up of large fragments of angular material cemented together by finer material. The term is normally applied to rocks formed by explosive volcanic activity and in these the cement is of ash of tuff.

Aggradation The process by which a land surface is built up by the deposition of debris and other solute material, especially by rivers. It usually results from the river's load bearing capacity being reduced. This may be as a consequence of an increase in the amount of debris supplied to the river, a loss of speed or volume of flow, or a rise in base level. Material is deposited on the stream bed, which accumulates with time.

Aggregate A mass or cluster of particles, most commonly used in reference to soils.

Agricultural region An extensive area of land that produces a distinctive pattern of agriculture different from that to be found in adjacent areas. A number of models have been developed to explain regional variations of agriculture.

Agriculture status of black soils Soils of this group are fairly fertile, giving, reasonable good yields despite continued cultivation without proper manuring. The content of water-soluble salts is high and hence the black soils are generally considered unsuitable for heavy irrigation. These soils are well suited to the cultivation of cotton, cereals and oil-seeds like linseed and safflower. Vegetables of different kinds and fruits of the citrus variety can also be grown successfully.

Agronomy *See* crop science.

A horizon A soil horizon consisting of well mixed organic material that lies near the ground surface. A horizon usually include a large proportion of humus, which gives the layer its dark coloration, and have had there many of their soluble minerals, removed by leaching (*i.e.* they are eluvial zones)

Air of soil The soil is not a continuous gaseous body as is the lower atmosphere air but is highly dispersed and absorbed by soil colloids. Moreover, an appreciable proportion of it is dissolved in the soil water. Again its humidity usually approaches 100 per cent, a condition especially favourable for the growth of fungi, bacterial and other soil organisms.

Air space of soil In a representative silt soil, it has thirty per cent of water space and 20 per cent of air space in the pores.

Alabaster A mineral form of gypsum ($CaSO_4.2H_2O$).

Algae of the soil Most algae are chlorophyll–bearing organisms and, as such must live at or very near the surface of the soil, if they are to survive in a habitat of this type. However, certain forms seem able to obtain their energy from organic matter and with such a capacity may exist within and below the surface horizon. The number of these, however, is not large. This, some algae live and function much like higher plants, while others in their nutritional habits are typical soil micro-organisms.

Alidade A surveying instrument used for drawing directions in plane table surveys. In its simplest form it consists of a ruler with raised sights a slot at one end and a hairline at the other. Having lined up the sights with a distant object the edge of the ruler can be used for drawing a direction line (ray).

Alkali A water soluble strong base. Strictly the term refers to the hydroxides of the alkali metals (group I) only, but in common usage it refers to any soluble base. Thus borax solution may be described as mildly alkaline.

Alkali flat A plain of sediment with a high proportion of alkali salts, characteristic of arid and semiarid regions. It is produced by the evaporation of a shallow mineral-rich lake, often in the dry season, which leaves behind a residue with a high concentration of sodium and potassium. In the wet season desert streams flowing into these areas produce play a lakes.

Alkaline rock An igneous rock that is rich in feldspar composed chiefly of sodium and potassium silicates. It contains 45–52 per cent silica and magnesium, is the second most abundant mineral. It is dark and heavy and forms the foundation of the ocean floor extending underneath the continents.

Alkali problem of soil When for any reason the drainage of an arid-region soil is impeded conditions become such as to favour the accumulation of soluble salts at or near the surface. This phenomenon is due to evaporation accompanied and abetted by an upward capillary adjustment of soil water, which gradually carries the excess salts onto the surface horizon. Such a concentration or accumulation is called alkali. A condition therefore results that often renders the soil practically useless for agricultural purpose.

Alkali reaction in soil A strong accumulation of the various salts in popularly called alkali as it usually has an alkaline reaction. It may be pointed out that dissolved salts in the sub-soils may not produce saline or alkali soils until irrigation is practised or the ground water rises to the surface and evaporates into the air, leaving behind a concentrated accumulation of dissolved salts.

Alkali soils These soils are intrazonal *i.e.* their outstanding characteristic, in such cases, the soluble salts, are a reflection of some local condition. And an impeded irrigation and drainage is the immediate cause of the excess of the soluble matter. They are completely devoid of vegetation. A white or greyish white incrustation can be seen on the surface during dry season.

Alkali soil management There are three general ways in which alkali lands can be handled in order to avoid at least partially the injurious effects of soluble salts. The first of these is eradication; the second is a conversion of some of the salts to less injurious forms; while the third which may be designated as control and in that the soil-cum-crop management procedures are employed.

Alkali spots Often it is seen that alkali is localised in small spots.

Allotropy The ability of certain elements to exist in more than one physical form. Sulphur and phosphorus are the most common examples. Allotropy is more common in groups IV, V, and VI of the periodic table than in other groups. *See* also enantiotropy, monotropy

Alloy A mixture of two or more metals (*e.g.* bronze or brass) or of a metal with small amounts of non-metals (*e.g.* steel). Alloys may be completely homogeneous mixtures or may contain small particles of one phase in the other phase.

Alluvial cone A cone-shaped accumulation of thick coarse alluvial material similar to, but steeper than, an alluvial fan. The landform is more typical of arid and semiarid areas where short-lived torrents carrying large loads are abruptly checked on entering a main valley or plain.

Alluvial deposits There are three kinds of general alluvial deposits. (1) Flood plains (2) alluvial fans and (3) deltas. These are rivers deposits, heterogeneous in composition; usually coarse and sandy in the upper reaches of stream and finer lower down; soil particles generally rounded; occur as narrow strips along river banks, or as 'alluvial fans'.

Alluvial fan A fan shaped landform composed of alluvium deposed where a constricted river enters a main valley or emerges from mountains onto a plain. The apex of the fan is located at a point at which the river is abruptly checked at the point. It divides to for several distributeries and deposit its load.

Alluvial flat A horizontal surface composed of alluvium deposited by a river in times of flood. Extensive flat areas of alluvium are known as alluvial plains.

Alluvial soil's suitability crop production Alluvial soils are suitable for the cultivation of wheat cotton, maize, millets, rice, oil-seeds, sugarcane and vegetables and fruits of many kinds, where irrigation facilities are high. In the heavy rainfall areas of Bengal, Bihar, Assam and eastern Uttar Pradesh, the soils are well suited to the growing of jute and paddy.

Alluvium Material deposited by a river along its course, this may be as cones, fans, deltas and flood plains and includes deposits on the stream bed in estuaries, and in lakes. The term is sometimes restricted to particles of silt size but in other definitions it may be extended to include sand and gravels. Alluvial deposits are agriculturally fertile although subject to flooding and can contain important minerals such as gold, platinum and diamonds.

Alpha-naphthol test (Molisch's test) A standard test for carbohydrates in solution. Molisch's regent, alphanaphthol in alcohol, is mixed with the test solution. Concentrated sulphuric acid is added and a violet ring at the junction of the two liquids indicates the presence of carbohydrates.

Alum A type of double salt. Alums are double sulphates obtained by crystallizing mixtures in the correct proportions.

Alumina *See* aluminium hydroxide.

Aluminate *See* aluminium hydroxide.

Aluminium A soft moderately reactive metal; the second element in group III of the periodic table. It has the electronic structure of neon plus three additional outer electrons. There are numerous minerals of aluminium; it is the most common metallic element in the Earth's Crust (8.1 per cent by weight) and the third in order of abundance. Commercially important minerals are bauxite (hydrated Al_2O_3), corundum (anhydrous Al_2O_3), cryolite (Na_3AlF_6), and clays and mica (aluminosilicates). The relationship between soil reaction and the activity of aluminium, iron and manganese seems to be somewhat more definite than the correlation with calcium and magnesium. When the pH of a mineral soil is low, appreciable amounts of aluminium as also of iron and manganese etc. may become toxic to certain lands. Aluminium reacts readily with the halogens; in the case of chlorine thin sheets will enflame. The fluoride has a high melting point (1290°C) and is ionic. The other halides are dimers in the vapour phase (two halogen bridges). Aluminium also forms a sulphide (Al_2S_3), nitride (AlN), and carbide (Al_4C), the latter two at extremely high temperatures. A number of very reactive aluminium alkyls are also known, some of which are important as polymerization catalysts. Symbol (Al; m.p. 659.8°C; b.p. 1800°C; r.d. 2.702; p.n. 13; R.A.M. 26.97.

Aluminium acetate (aluminium ethanoate, $Al(OCCH_3)_3$ A white solid soluble in water. It is usually obtained as the dibasic salt, basic aluminium ethanoate $[Al(OH)(C_2H_2O_2)_2]$.

Aluminium bromide ($AlBr_3$) A white solid soluble in water and many organic solvents.

Aluminium chloride ($AlCl_3$) A white covalent solid that fumes in most air

$$AlCl_3 + 3H_2O - Al(OH)_3 + 3HCl$$

It is prepared by heating aluminium in dry chlorine or dry hydrogen chloride.

Aluminium fluoride (AlF_3) A white crystalline solid that is slightly soluble in water but insoluble in most organic solvents. Its primary use is as an additive to the cryolite (Na_3AlF_6) electrolyte in the production of aluminium.

Aluminium hydroxide [$Al(OH)_3$] A white powder prepared as a colourless gelatinous precipitate by adding ammonia solution or a small amount of sodium hydroxide solution to a solution of an aluminium salt. It is an amphoteric hydroxide. It is used as a foaming agent in fire extinguishers and as a mordant in dyeing.

Aluminium nitrate [$Al(NO_3)_3.9H_2O$] A hydrated white crystalline solid prepard by dissolving freshly prepared aluminium hydroxide in nitric acid. It cannot be prepared by the action of dilute nitric acid on aluminium since the metal is rendered passive.

Aluminium oxide (alumina, Al_2O_3) A white powder that is almost insoluble in water. Because of its amphoteric nature it will react with both acids and alkalis. Aluminium oxide occurs naturally as bauxite, corundum, and white sapphire.

Aluminium on phosphorus If the superphosphate added to the soils that are quite acidic, say pH 5 or below, the available phosphates react with aluminium and iron, which are active at the pH range. The reversion products are even more insoluble and in available than the Calcium Phosphates formed when the pH is above 7.

Aluminium potassium sulphate *See* alum.

Aluminium toxicity When the pH of a mineral soil is low, appreciable amount of aluminium and also of iron and manganese, are active, so much so in fact, that the aluminium may become extremely toxic to certain plants.

Amalgam An alloy of mercury with one or more other metals. Amalgams may be liquid or solid. An amalgam of sodium (Na/Hg) with water is used as a source of nascent hydrogen.

Amino acids Derivatives of carboxylic acids in which a hydrogen atom in an aliphatic acid has been replaced by an amino group. Thus from ethanoic acid the amino-acid 2-aminoethanoic acid (glycine) is formed.

Ammonia It is a gas containing about 80 per cent of nitrogen (NH_3). Under suitable conditions of temperature and pressure, it becomes liquid (anhydrous ammonia). Another form, 'aqueous ammonia', results from the absorption of ammonia gas in waster in which it is soluble. Ammonia is used as a fertilizer in both these forms. Anhydrous ammonia can be applied by introducing it into irrigation water, or directly into the soil from special containers, which make its use rather expensive. Its possibilities as manure for paddy, sugarcane and cotton are being investigated at Bangalore in Mysore State.

Ammonical nitrogen Considerable amounts are appropriated by some of the ammonifers themselves or by other soil organisms capable of using this type of compound. Mature plants such as certain grasses, forest trees and their undergrowth apparently are forced to use ammonical nitrogen because the soils in which they grow often do not support ready nitrification. Also, when animal and plant synthesis temporarily is satisfied, and if by-product ammonium still remains unused, nitrogen in this form may go in third direction.

Ammonification Ammonium salts are formed only when nitrogenous bodies are involved. They are the first really simple nitrogen compounds that appear as a process of microbial digestion goes on. Amino acids and similar nitrogenous materials readily yield ammonium compounds by enzymic hydrolysis. The transformation is the ammonification.

Ammonium anhydrous Contains 82 per cent nitrogen.

Ammonium chloride (sal ammoniac, NH_4Cl) A white crystalline solid with a characteristic saline taste. It is very soluble in water (37g per 100g of water at 20°C). Ammonium chloride can be manufactured by the action of ammonia on hydrochloride acid. It is a white crystalline compound possessing good physical condition and containing 26 per cent of ammoniacal nitrogen. It is extensively used on paddy in Japan. In India, it is used largely for industrial purposes. In general, it is similar to ammonium sulphate in action. It is usually not recommended for tomatoes, tobacco and such other craps that may be injured by chlorine.

Ammonium fertilizers Many ammonium salts such as nitrate of soda, ammonium sulphate and others are commonly used as fertilizers in crop production.

Ammonium ion The NH_4^+, formed by coordination of NH_3 to H^+. *See* also quaternary ammonium compound.

Ammonium nitrate (NH_4NO_3) It is white crystalline salt, containing 33 to 35 per cent nitrogen, half as nitrate nitrogen and half in the ammonium form; the latter cannot be readily leached from the soil. This fertilizer is quick acting but highly hygroscopic and not fit for storage. Granulation of the material and a light coating of granules by oil reduces hygroscopicity to some extent. I has an acidulating affect in soil. Under certain conditions, it has explosive qualities and, therefore, it should be handled cautiously. Highly soluble in water 871 g per 100 g of water at 100°C). It is usually manufactured by the action of ammonia on nitric acid. It is used in the manufacture of explosives and, because of its high nitrogen content as fertilizer.

Ammonium phosphate This fertilizer is soluble in water and contains 11 per cent nitrogen and 48 per cent phosphoric acid.

Ammonium salts Under humid-temperature climate an average of about 4½ lbs. If ammoniacal and 1½ lbs. of nitrate nitrogen fall on each acre in land yearly through rain water.

Ammonium sulphate $(NH_4)_2SO_4$ A colourless crystalline solid that is soluble in water. When heated carefully it gives ammonium hydrogensulphate, which on stronger heating, yields nitrogen, ammonia, sulphur dioxide, and water. Ammonium sulphate is manufactured by the action of ammonia on sulphuric acid. It is the most important ammonium salt because of its widespread use as a fertilizer. Its only drawback as a fertilizer is that it tends to leave its acidic residue in the soil. It is the most widely used fertilizer. It is a white crystalline salt, containing 20 to 21 per cent ammonical nitrogen. It is easy to handle and it stores well in dry conditions. During the rainy season, it sometimes forms lumps. These should be powdered before use. Being soluble in water, it acts quickly. But despite its high solubility, its nitrogen is not readily lost in drainage because the ammonium ion is retained by the soil particles. Ammonium sulphate can be applied prior to sowing at moving time, or as a top dressing to the growing crop. It should not be applied with or too close to the seed, because in concentrated forms, it affects seed germination very adversely.

Ammonium sulphate nitrate This mixture of ammonium nitrate and ammonium sulphate is available in white crystalline form or as granules of dirty white colour. This fertilizer contains 26 per cent nitrogen, three-fourths being in the ammoniacal form and the rest (6.5 per cent) as nitrate nitrogen. It is non-explosive and not as deliquescent as ammonium nitrate. It is readily soluble in water and very quick action. Its keeping quality is good and it is useful for all crops. Its acid effect on soils is only one-half of that of ammonium sulphate. It can be applied before sowing, time or as a top-dressing but it should not be applied along with seed.

Ammonium superphosphate It has 4 per cent nitrogen and 18 per cent phosphorus.

Ammophos A Contains 11 per cent nitrogen and 45–48 per cent phosphorus.

Ammophos B Contains 16.5 per cent nitrogen and 20 per cent phosphorus.

Ammo-phos-$NH_4H_2PO_4$ Contains available phosphoric acid and nitrogen.

Amount of elements in soils The total amount of elements contained in soils depends partly on the nature of the rocks from which they are formed and partly on their age and the extent to which soluble products have been leached away. The chemical composition of different horizons of a soil also shows a good deal of variation.

Amount of essential elements in soil Although soils vary greatly in their chemical composition it is possible the percentage range within which the primary nutrients are ordinarily found when a number of surface soils are to be considered.

Amount of fertilizers to apply The kind and amount of the fertilizer applied should, as nearly as possible, meet the changing conditions in the soil. The reversion of phosphoric acid and decomposition of organic ammoniates should be given due allowance to be made for such contingencies when deciding on the kind and amount of fertilizer to apply.

Amount of hygroscopic water in soils For a sandy loam containing a little of organic matter, the hygroscopic coefficient may be as low as 1 to 2 per cent by weight, while in heavy soil with organic matter of 4 to 5 per cent it may rise as high as 12 to 15 per cent. This inactive water is large in some soils.

Amount of lime to apply The soil pH is important for this purpose as it gives some idea as to the percentage base saturation of the soil and how acute is the need for lime. The texture and organic matter are important since they are indicative of the absorption capacity of the soil and the strength of buffering.

Amount of nitrogen fixed by legumes In an experiment carried at Ithaca, New York, gave the following results showing that different plants fix different quantities of nitrogen in soil.

Alfalfa	100	Soyabean	42
Sweet clover	67	Hairy vetch	27
Red clover	56	Field beans	23
Alsike clover	56	Field peas	19

Amount of nitrogen fixed by nodule bacteria This depends on many factors. The condition of the soil, especially as to aeration, drainage, moisture, and the amount of active calcium. Some soils are in such condition as to harbour only a meagre, non-virulent flora of legume bacteria. Besides, there is the nature of host plant, some facilitate more fixation than others.

Amount of primary nutrient Soils vary in their chemical compositions it is possible to indicate the percentage range within which the primary nutrients are ordinarily found when a surface soil is considered.

Amount of soil granulated In well-granulated soils, 70–80 per cent of the total mass may exhibit a crumb-structure type of aggregation, although 40–50 per cent is very common.

Ampere symbol–A The SI base unit of electric current, defined as the constant current that, maintained in two straight parallel infinite conductors of negligible circular cross section placed on metre apart vacuum, would produce a force between the conductors of 2×10^{-7} newton per metre.

Amphoteric A material that can display both acidic and basic properties. The term is most commonly applied to the oxides and hydroxides of metals that can form both cations and complex anions. For example, zinc oxide dissolves in acids to form zinc salts and also dissolves in alkalis to form zincates, $[Zn(OH)_4]^2$.

Anemometer An instrument that measures the velocity and often the direction of the wind. The most common is the cup-anemometer in which symmetrically mounted cups rotate on a vertical axis, the speed of rotation indicating the strength of wind. Other include the pressure tube anemometer in which the indication is the pressure built up in a tube that is closed at one end; the pressure plate anemometer, which measures the pressure exerted by the wind in a metal plate; and the hot wire anemometer, in which the current required to keep the temperature of the platinum wire constant indicates the wind speed.

Angstrom symbol–A Unit of length defined as 10^{10} metre. The angstrom is sometimes still used for expressing wavelengths of light or ultraviolet radiation or for the sizes of molecules, although the manometer is preferred. It is equal to 0.0000001 millimetre.

Anhydride A compound formed by removing water from another compound. Many non-metal oxides are anhydrides of acids; for example CO_2 is the anhydride of H_2CO_3'SO_3 is the anhydride of H_2O from two carboxylic-acid groups, giving compounds with functional group -CO.O.CO-. These form a class of organic compounds called acid anhydrides (or acyl anhydrides). *See* acid anhydride.

Anhydrite *See* calcium sulphate.

Anion A negatively charged ion, formed by addition of electrons to atoms or molecules. In electrolysis anions are attracted to the positive electrode (the anode). Compare cation.

Annular drainage A pattern of drainage in which steams follow rough circular paths. Such patterns are usually produced on domed structures where the rivers follow the outcrops of the weaker beds of rock.

Anode In electrolysis, the electrode that is at a positive potential with respect to the cathode. In any electrical system, such as a discharge tube or electronic device, the anode is the terminal at which electrons flow out of the system.

Anodizing An industrial process for protecting aluminium with an oxide layer formed in an electrolytic cell containing an oxidizing acid [*e.g.* sulphuric (VI) acid]. The layer of Al_2O_3 is porous and can be coloured with certain dyes.

Anthracite A type of land brittle shiny coal. It is formed when bituminous coal is subjected to a great pressure resulting burial and earth movement accompanied by a slight increase in temperature which drives of nearly all the volatile material.

Anticline A rock structure in which the beds of rock are folded into an arch *i.e.* the beds dip onwards from the crest, the oldest rocks occurring in the core. Anticline alternate with syncline and are formed through compression of the rock layers.

Antimony A toxic element existing in two allotropic forms; the most stable in a brittle silvery metal. Arsenic belongs to group V of the periodic table. It is found in many minerals, principally stibnite (Sb_2S_3). It is used in alloys and semiconductor devices. Symbol: sb; m.p. 630.7°C; b.p. 1640°C; rd. 6.69; p.n. 51; r.a.m. 121.7.

Apatite A naturally occurring phosphate of calcium. $CaF_2Ca_3(PO_4)_3$. It is one of the soil minerals, $Ca_5(PO_4)_3.(C1,F)$. It is a source of phosphorus in a mineral soil, it contains 30–60 per cent phosphorous.

Apparent weight of a body in a moving lift If a body of mass m is placed in a lift moving upwards with an acceleration of a, then on lift is weight = R.–mg = ma; R = mg + ma = mg(1 + a/g), where R is also called the apparent weight of the body.

Applied climatology *See* crop science.

Aqua regia A mixture of concentrated nitric acid and three to four parts of hydrochloric acid. It dissolves all metals including gold hence the name. The mixture contains chlorine and NaCl.

Aqueous Describing a solution in water.

Aquifer (aquafer) A bed of rock that holds water and also allows water to flow freely through it. *See* also artesian basin.

Arable land Farm land that is suitable for ploughing or cultivation or is actually ploughed and cropped. It includes land that is temporarily fallow or cultivated with rotation grasses.

Arcuate delta A fan-shaped delta with its rounded margin extending into the sea. The fan shape is a result of the division and subdivision of channels. The classic example of an arcuate delta is the Nile Delta. *See* also delta.

Area 1. The extent of a particular expanse of land or surface enclosed within a specified boundary; it is measured in square units, *e.g.* square kilometres (km^2).

2. Any expanse of land forming a region, district or locality.

Arenaceous Denoting a medium grained sedimentary rock with a sandy texture, such as sandstone and coarse silt-stone. Arenaceous rocks are also often collectively known as sandstones. They have a grain size (on the Wentworth scale) 1/16 mm to 2 mm.

Arene An organic compound containing a benzene ring; *i.e.* an aromatic hydro-carbon or derivative.

Argillaceous It denotes a fine grained sedimentary rock composed largely of clay particles with a grain size of less than 1/16 mm (Wentworth scale). Argillaceous rocks include silt, stones, clays, mud-stone, shales and marls.

Argon An inert colourless odourless monatomic element of the rare-gas group. It forms 0.93 per cent by volume of air. Argon is used to provide an inert atmosphere in electric and fluorescent lights and in welding.

Arid Denoting any climate or region in which the rainfall is insufficient or barely sufficient to support vegetation. It is sometimes defined as an area having an average annual rainfall of less than 250 mm but this is not a generally accepted definition.

Arid and desert soils The soil of the whole tract is a vast sandy plain including isolated hills and rock out-crop at places. The soils improve in fertility towards east and north-east. Usually the soils are saline or alkaline with unfavourable physical condition and high pH value. Water is scarce and occurs at great depths. The phosphate content of the aeolian sand compare favourably with some of the alluvial sols, P_2O_5 ranging from 0.05 to 1 per cent. Total nitrogen is low, mostly ranging between 0.02 to 0.07 per cent. The pH of the soils and sand varies between 7.2 and 9.2. The salt content is not in toxic doses. The soils and sand are calcareous in nature. Nitrogen occurs in the form of nitrates which contribute to soil fertility.

Aridic group of soils *See* arid soils. Biologically, organisms may be active at greater depths in zonal aridic than in zonal humid soils because of loose structure and good aeration of the former. But the intensity of biological activity in arid soils is very largely governed by moisture.

Arsenic A toxic metalloid element existing in several allotropic forms; the most stable is a brittle grey metal. It belongs to group V of the periodic table.

Arsine (arsenic hydride, AsH_3) A poisonous colourless gas with an unpleasant smell. It decomposes to arsenic and hydrogen at 230°C. It is produced in the analysis for arsenic (Marsh's test).

Artesian basin A geological structure in which the beds dip toward a central point forming a basin with a permeable player of rock (acquifer) sandwiched between impermeable beds. Water seeping into the aquifer is trapped between the impermeable layer where it accumulates and eventually the rock becomes saturated.

Aryl group An organic group derived by removing a hydrogen atom from an aromatic hydrocarbon or derivative.

Asbestos A fibrous mineral that occurs in several forms, the most important commercially being chrysotile. It can be spun and woven into cloth or compressed into boards but its most important properties are its resistance to heat and decay; for making fireproof clothing and in the construction industry. The major producers are Canada, South Africa, Zimbabwe and Italy.

Ash Fine-grained particles of lava thrown out during volcanic eruptions.

Aspect The direction in which a slope faces. The amount of sunshine received by a slope will vary according to its aspect. This has strong influence on the development of soils, vegetation, and drainage and on the location of settlements.

Assembling of aggregate It is very difficult to explain as what draws and hold the individual granule together once the proper contacts have been made and the aggregate is assembled. Evidently, the chemical and physical inter-relationship of clay, hums and water play the principal roles and also the electrical condition of these components fundamentally involved, as well as dehydration precipitation and cementing effects also matter a lot.

Atmometer *See* evapometer

Atmosphere The layer of gases enveloping the earth. It is composed of nitrogen (78 per cent), oxygen (20.95 per cent), argon (0.935 per cent) and carbon dioxide (0.03 per cent) together with small proportions of other gases and varying amounts water vapour. Some solid particles from such sources as volcanic eruptions, sea salt and wind blown dust are also contained in the atmosphere. Nearly 90 per cent of the atmosphere lies within 29 kms. of the surface of the earth although its outer edge lies at about 10,000 kms. There are several layers in the atmosphere, the lowest of which is troposphere in which most meteorological phenomena occur. Above this is the stratosphere, mesosphere and thermosphere.

Atmospheric pressure The pressure exerted by the atmosphere as a result of its weight above a unit area of the earth surface. It is expressed in milibars (mb) and measured with a mercury barometer. The average atmospheric pressure at sea level is 1013.25 mb. Seal level values can range from extremes of about 890 mb to 1060 mb in hurricanes to 1060 mb in anticyclones. Atmospheric pressure decreases logarithmically with height.

Attrition The wearing down of rock by other rock particles while being transported by water, wind or ice. This process results in the particles becoming gradually smaller, smoother and more rounded such as beach pebbles.

Augite This made up of silicates of calcium, magnesium, iron and sodium in varying proportions.

Autotrophs *See* plant science.

Availability of inputs Basically this means availability of water, fertilizers and improved genetic materials. An important problem is not only to increase the yields of crops per unit area but also per unit input.

Available forms or compounds for plants For example, if phosphorus is given to the soil in the form of tri-calcium phosphate, it cannot be used by the plants in this form. To be made available to the plants, it should be given as mono-calcium phosphate [$Ca(H_2PO_4)_2$].

Available phosphoric acid The term available actually refers to those phosphates that readily stimulate plant growth when properly used as fertilizers. The phosphates which are less effective are rated unavailable.

Available water of soil Three divisions of soil water may be (1) unavailable (2) desirably available (3) and superfluous.

Avogadro's law The principle that equal volumes of all gases at the same temperature and pressure contain equal numbers of molecules. It is often called Avogadro's hypothesis. It is strictly true only in ideal gases.

Azine An organic heterocyclic compound that has a hexagonal ring containing carbon and nitrogen atoms. Pyridine is the simplest example.

Azo compund A type of organic compound of the general formula RN : NR, where R and R are aromatic groups. Azo compounds can be formed by coupling a diazonium compound with an aromatic phenol or amine.

Azofication There exist in the soils certain organisms, using the organic matter as a source of energy, that are able to obtain elemental nitrogen from the soil, air. This nitrogen is incorporated in the bodies of the organisms and is left in the form of proteins and related compounds upon their death. The process is called azofication, because of the azotobacter group of bacteria to which is ascribed a major part of the fixation.

Azonal groups of soils These soils are without characteristics. Most alluvial and colluvial soils belong to this group. Also called skeletal soil. A soil in which the profile is undeveloped and little or no differentiation into separate soil horizons has taken place. A zonal soils are often young, *i.e.* soil forming processes have had insufficient time to operate, as on recently deposited alluvium, reclaimed land, windblown deposits, or a new lava flow. B are rocky surfaces such as screes are a zonal soils. Compare intrazonal soil, zonal soil.

B

Backward reading vernier If n V.D. are equal to (n–L) main scale division, it is called, Backward Reading Vernier.

Bacteria of soil It is the most important class of the soil micro-organisms. These are single celled organisms, the simplest and smallest forms of life. They multiply by elongating and dividing into half. They are therefore sometimes called fission fungi.

Bacterial activity in soil One of the most important activity of bacteria is to convert the comparatively insoluble complex organic compounds of nitrogen into ammonia; another (name Nitrosomonas) changes ammonia to nitrite; and a third species (Nitrobacter) transforms nitrites into nitrates. The latter two are collectively known as nitrifying bacteria and the process as a whole is called nitrification. Though found in almost all cultivated soils, the nitrifying bacterial (and consequently nitrate formation) are generally confined to their top 10 to 2 inches. The organisms work best at temperatures between 25° to 38°C, and under conditions of thorough tillage, good soil aeration, neutral reaction and a moisture content of about 60 per cent of the soil's water holding capacity.

Bacterial colonies in soil in soil, bacteria exist as mats, clumps, and filaments called colonies, on and around the soil particles wherever food and other conditions are favourable.

Balance of nature *See* ecological regulation.

Barium A dense, low-melting reactive metal; the fifth member of group II of the periodic table and a typical alkaline-earth element. Barium metal is used as a 'getter' *i.e.* a compound added to a system to seek out the last traces of oxygen; and as an alloy constituent for certain bearing metals. Notable differences in the chemistry of barium from the rest of the group are:

1. The much higher stability of the carbonate.
2. The formation of the peroxide below 800°C. Barium peroxide decomposes on strong heating to give oxygen and barium oxide

$$BaO_2 = BaO + O$$

The equilibrium was the basis of the totally obsolete process for the commercial production of oxygen. Barium is also notable for the very low solubility of the sulphate, which permits its application to gravimetric analysis for either barium or sulphate. Barium compounds give a characteristic green colour to flames which is used in qualitative analysis. Barium salts are all highly toxic with the exception of the most insoluble materials. Metallic barium has the body centred cubic structure. Symbol Ba; m.p. 714°C; b.p. 1537°C; r.d. 3.51; p.n. 56; r.a.m. 137.34.

Barium bicarbonate ($BaCO_3$) A white insoluble salt that occurs naturally as the mineral witherite. Barium carbonate can be readily precipitated by adding an alkali carbonatre to a barium salt solution. On heating it decomposes reversibly with the formation of the oxide and carbon dioxide

$$BaCO_3 = BaO + CO_2$$

It is used as a rat poison.

Barium chloride ($BaCl_2$) A white solid that can be prepared by dissolving barium carbonate in hydrochloric acid and crystallizing out the dihydrate ($BaCl_2 2H_2O$). Barium chloride is used as the electrolyte in the extraction of barium, as a rat poison, and in the leather industry.

Barium hydroxide [baryta, $Ba(OH_2)$] A white solid usually obtained as the octahydrate, $Ba(OH_2)$ $8H_2$. Barium hydroxide is the most soluble of the group IIA hydroxides and can be used in volumetric analysis for the estimation of weak acids using phenolphthalein as an indicator.

Barium oxide (BaO) A white powder prepared by heating barium in oxygen or by thermal decomposition of barium carbonate. It has been used in the manufacture of lubricating-oil additives.

Barrel The standard unit of volumetric measurement of petroleum or crude oil, equal to 159 litres (42 US gallons or 35 Imperial gallons).

Barren lands (barrens) Flat open lands with vegetation limited to small trees and sparse shrubs. The name is applied to the tundra plains of 'N Canada,' which are also referred to as the Arctic prairies.

Basal sapping The erosion and removal of debris at the foot of a slope. This may result from a number of different processes.

Basalt A fine-grained dark igneous rock of basic composition, comprising over 90 per cent of all volcanic rocks. The dark colouring is due to its chemical composition; it consists chiefly of plagioclase feldspar and pyroxene, with magnetite, apatite and olivine as accessory minerals.

Base-catalysed reaction A reaction catalysed by bases. Typical base-catalysed reactions are the Claisen condensation and the aldol reaction, in which the first step is abstraction of a proton to give a carbanion.

Base exchange It is generally assumed that adsorbed nutrients are rather readily available both to higher plants and to micro-organisms. As how is the transfer from soil colloid to soil organisms doubtedly facilitates the availability of nutrients held in this way. One of the common procedures is to displace the original adsorbed cation by treating the sample with an ammonia or barium acetate solution of a given pH. The capacity of clay or soil to adsorb ammonia or barium ion under controlled conditions is taken as a measure of exchange capacity.

Base exchange and loss of time After the production of carbonic acid in the soil, the interchange of hydrogen and calcium is brought about instantaneously in chemical equivalents. And if the soluble bicarbonate is removed by drainage, the movement of the lime will be continuously outward. This is very common in humid soils.

Base exchange capacity of clay It is another term for the relative adsorption power of a clay, the method of expressing its magnitude is by the unit milliequivalent.

Base exchange due to potassium chloride A soil fairly high in replaceable calcium, is if treated with a liberal application of potassium chloride, a common fertilizer salt. Some of the added potassium pushes its way into the colloidal complex and forces out its equivalent of calcium and other ions. These appear in the soil solution.

Base exchange in a normal cultivated mineral soil They afford a more tangible idea of base-exchange capacity and other related characteristics. The figures so obtained may be modified to suggest the exchange condition of a corresponding arid-region surface soil by lowering the adsorbed hydrogen and increasing the metallic cations proportionately until the percentage base saturation reaches 90 or 100. The pH approaches 7 or may even exceed this figure if carbonates are present.

Base exchange in organic colloids In respect of base exchange capacity the contrast of humus with clay is striking. The collodial silicate minerals vary in respect to their exchange capacity.

Base exchange in organic soils Because of the large amount of exchangeable calcium and magnesium present in the organic soils and the marked evolution of carbon dioxide, if drainage is alright and under cultivation, base exchange is as active as in mineral soils.

Base exchange of soils due to bicarbonates The presence of carbonic acid (H_2CO_3) in the soil solution is a must. The hydrogen thus generated is exceedingly active and ends to replace the exchangeable calcium of the colloidal complex. This occurs not only because of mass action effects but also because ionic hydrogen is adsorbed more strongly than is the ionic calcium.

Base exchange of calcium When much of the potassium is transferred due to base exchange after water and carbon dioxide have produced soluble potassium compounds, a great deal of calcium becomes mobile, by a simple exchange of hydrogen ions for those of calcium. This replacement or ionic exchange is one of the most types of reactions in the soil. Because of this situation, soils in general though containing in general considerably more potassium than calcium, liberate the latter element much more lavishly. This has a direct bearing on calcium nutrition, soil acidity, liming and other practical things.

Base exchange of silicate clays Clays differ markedly in their base exchange capacity and properties. Hence soils will differ widely in the important capacity, not only because they posses different amounts of clay but also because the clay complex is so variable.

Base exchange reversibility The reversibility of any exchange reaction is highly probable. If lime is applied to an acid soil, *i.e.* in one in which the colloidal complex is definitely dominated by hydrogen, a replacement, the reverse of active calcium reaction.

Base exchange significance It is most important chemical reaction next to photosynthesis in the whole field of agriculture. Considering the influence of ionic exchange on the physicochemical conditions of soils, on the availability of nutrients to micro-organisms and higher plants, on the loss of soluble constituents in drainage, and on the effectiveness of line and commercial fertilizers, these all signify the usefulness of this phenomenon.

Base holding capacity of soils It also depends on colloid composition. In general, soils of high base holding capacity are those in which colloids have a high silica-sesquioxide ratio. They have also usually a high exchangeable base content. Such soils are generally fertile.

Base metal A metal such as iron, lead, or copper, distinguished from a precious metal such as gold or silver.

Basic aluminium ethanoate *See* aluminium acetate.

Basic lava A lava with a low viscosity and low melting point that flows readily, often for great distances. It is rich in iron magnesium, and other metallic elements, but poor in silica. It forms shield valcanoes, such as those of Hawiian islands, which are wide spreading domes with very gentle slopes.

Basic rock An igneous rock that is quartz free and contains a low percentage of silica (45–55 per cent by weight). It is composite of minerals such as calcium feldspars. Rock of this type form the asthenosphere. Such rocks include basalt, dolorite and gabbro.

Basic salt A compound intermediate between a normal salt and hydroxide or oxide. The term is often restricted to hydroxyhalides [such as $Pb(OH)Cl$, $Mg_2(OH)_3Cl_2$, $4H_2O$, and $Zn(OH)\ F$] and hydroxy-oxy salts (for example, $2PbCO_3$, $Pb(OH)_2$, $Cu(OH)_2$. $CuCl_2$ $Cu(OH)_2$ $CuCO_3$). The hydroxyhalides are essentially closely packed assemblies of OH^- with metal or halide ions in the octahedral holes. The hydroxy-oxy salts have more complex structures than those implied by the formulae.

Basic slag It contains 16–20 per cent phosphoric acid. It should be ground fine before giving it to soil.

It is more useful in acidic soils.

Basin 1. A large-scale depression in the surface of the Earth, produced by erosion or by the underlying geological structure. 2. In geology, a structural downfold in the Earth's crust in which the rocks dip towards the same central point. 3. The area drained by a single river system, *i.e.* a drainage basin.

Bauxite A clay-like amorphous material, the chief ore of aluminium. It is formed when silica is removed from feldspars by chemical weathering. The remaining aluminium hydroxide is insoluble and collects in rock-lime layers at or near the surface as the parent rock weathers away.

Beaufort wind scale A scale of wind force. It was devised in 1805 by Admiral Sir Francis Beaufort and modified in 1926. The scale which ranges from 0 to 12 is based on easily observable indicators of wind strength such smoke, tree movement and damage incurred.

Beckmann thermometer A type of thermometer designed to measure differences in temperature rather than scale degrees.

Bed 1. In geology, a single layer or stratum of sedimentary rock. The bed of rock is separated from the beds above and below by distinct surfaces known as bedding planes. 2. The bottom of a river channel, lake or sea.

Bedding The arrangement of sedimentary rocks in layers. *See* also bedding plane, stratum.

Bedrock The solid rock lying beneath soil and weathered material. It is usually only exposed as an outcrop on steep slopes where no soil is present.

Belt A broad strip of land characterised by one dominant feature, such as a crop or climatic type. Examples are the Corn Belt and Cotton Belt.

Bench A narrow terrace-like landform, characterised by a flat surface with a steep slope at the back. Benches may be erosional in origin, the most common being produced by lakes or rivers that have experienced a change in the base level. They also result from structural processes, *e.g.* step faulting, and artificial excavations, such as quarrying and open cast mining.

Bench mark A mark made by surveyors to record a point of known position and height above mean sea level that can be referred to subsequently when further surveys are carried out. The mark is cut into a permanent feature such as a rock or stone facing of a building. It consists of a broad arrow with a horizontal bar through its apex. The arrow points upwards unless the location is below sea level in which case it is inverted. Bench marks are shown Ordnance Surveys (OS) maps by the letters BM and the height.

Bench terracing This consists of a series of platform having suitable vertical drops along contours or on suitably graded lines across the generally slope of the land.

Benzaldehyde *See* benezene carbaldehyde.

Benzene (C_6H_6) A colourless liquid hydrocarbon with a characteristic odour. Benzene is a highly toxic compound and continued inhalation of the vapour is harmful. It was originally isolated from coal tar and for many years this was the principal source of the compound. Contemporary manufacture is from hexane; petroleum vapour is passed over platinum at 500°C and at 10 atomspheres pressure.

Bergius process A process formerly used for making hydrocarbon fuels from coal. A powdered mixture of coal, heavy oil, and a catalyst was heated with a hydrogen at high pressure.

Beryllium A light metallic element, similar to aluminium but somewhat harder, the first element in group II of the periodic table. It has the electronic configuration of beryllium with two additional outer 2s electrons.

Beryllium sulphate ($BeSO_4$) An insoluble salt prepared by the reaction of beryllium oxide with concentrated sulphuric acid. On heating, it breaks down to give the oxide. When heated in the presence of water, it readily dissolves to give an acidic solution.

Bessemer process A method for making steel from pig iron, invented by Sir Henry Bessemer in 1985. The process takes place in a Bessemer converter, a large steel vessel. Air is blown through molten pig iron and impurities in the iron (carbon, silicon, and manganese) are oxidized. The impurities from slag, which can be drawn off. Phosphorous is removed from the iron by reaction with the basic refractory lining of the converter.

B horizon A soil horizon occurring in many soil profiles between the A and C horizons from which it is usually well-distinguished. The B horizon is often known as the subsoil. The B horizon receives material leached down from the A horizon, *e.g.* fine clay particles or ferric oxides. This material is sometimes concentrated in the form of a pan. B horizons are sometimes called illuviated horizons.

Bial's reagent A mixture of orcinol in concentrated hydrochloric acid with 10 per cent iron (III) chloride added, used in testing for pentoses. A positive reaction is shown by a green coloration produced when the test material is boiled with the reagent.

Biomolecular Describing a reaction or step that involves two molecules.

Binary compound A chemical compound formed from two elements; *e.g.* Fe_2O_3 or NaCl.

Biological agencies in soil formation Such agencies also assume a considerable importance within soil itself as the action of plant roots and earthworms can affect.

Biological method of soil reclamation By growing of salt or alkali tolerant crop plants, such as paddy, barley, sugarcane and castor helps in gradual reduction of salt concentration.

Biological nature of soil Generally in any mineral soil, there is much available energy in organic form, harbours a varied population of living organisms. Nature is most lavish in this respect. Indeed the life of the soil plays such a prominent and indispensable role in the changes, constantly occurring, that no discussion of the organic matter is complete without its consideration. Both animals and plants are abundantly represented in soil. Some of the organisms are large and others are very small. The whole gamut from the larger rodents, worms and insects to the tiniest bacteria commonly occurs in the normal soil.

Bismuth A brittle pinkish metallic element belonging to group of the periodic table. It occurs native and in the ores Bi_2S_3 and Bi_2 and Bi_2O_3.

Bittern The liquid that is left after sodium chloride has been crystallized from sea water.

Bitumen A mixture of solid or semisolid hydrocarbons obtained from coal, oil, etc.

Bituminous coal A soft black coal, containing 48-83 per cent carbon. It is the most common type of coal and is often known as household coal. Various types of bituminous coal exist and are named after the uses to which they are most frequently put, *e.g.* cooking coal and gas coal.

Black alkali Carbonates of the metals, particularly sodium carbonate, dissolve organic matter from the soil, thus giving a dark colour to the solution and so the incrustation, for this reason, alkali containing large quantities, especially of carbonate of soda, is called black alkali.

Black alkali soil *See* solonetz.

Black band ironstone A layer of carbonaceous ironstone that occurs in the Coal Measures. It is the most valuable of the iron ores of the Coal Measures. As it has a coal content of 10–20 per cent and virtually no clayey material it can be smelted economically.

Black cotton soil *See* regur. The group derives its name from the black colour of its soils and their eminent suitability to the growing of cotton.

Black earth *See* chernozem.

Black lead *See* graphite.

Black fixe *See* barium sulphate.

Blanket bog A bog that forms under conditions of high rainfall and high humidity. It usually occurs as a continuous cover over the land, broken only by steep slopes and rocky outcrops. *See* also bog, swamp.

Blast furnace A furnace for producing pig iron from iron ore. A tall steel container lined with heat-resistant bricks is filled from the top with iron ore, coke, and limestone. A hot blast of air is blown into the base of the furnace and the coke bums melting the ore. This settles at the bottom of the furnace to be drawn off every three to five hours. The furnace is continuously topped up until the lining needs renewal or demand for the pig iron drops.

Bleaching powder A white solid that can be regarded as a mixture of calcium chlorate (calcium hypochlorite) calcium chloride, and calcium hydroxide. It is prepared on a large scale by passing a current of chlorine through a tilted cylinder down which is passed calcium hydroxide.

Block disintegration A type of mechanical weathering resulting chiefly from frost action. When water present in the pores and joints in rocks freezes ice crystals grow which exert great pressure on the rock. This pressure causes the rock to fragment and crack along line of weakness, producing a land surface littered with large angular boulders, usually referred to as felsenmeer. *See* also weathering.

Block faulting The breaking up of an area of the Earth's crust into blocks, which may be raised or lowered. Raised blocks are called horsts and lowered blocks are called grabens. *See* also fault.

Blocky Here the aggregates are in the shape of irregular blocks with rather edges. These blocks are firm and solid and range in different soils from a friction of an inch in diameter to 2 or 3 as maximum. A blocky structure is commonly found in the B horizon of soils.

Blossom showers Mango showers the rain showers that occur from March to May in the monsoon region of S.E. Asia.

Blue green algae Most of the species of blue green algae, possessing the capacity of fix atmospheric nitrogen belong to the order Nostocales and Stigonematales.

Bog An area of wet spongy ground consisting of decomposing moss and other vegetation. It often forms with the growth of moss, especially sphagnum, on the surface of a shallow pond or lake.

Decaying moss and other vegetable matter will gradually accumulate, infilling the pond, to produce a quaking bog. *See* plant science.

Bog lime In many organic soils are under-laid at varying depths by a soft impure calcium carbonate called bog lime or marl.

Boiling The process by which a liquid is converted into a gas or vapour by heating at its boiling point. At this temperature the vapour pressure of the liquid is equal to the external pressure, and bubbles of vapour can form within the liquid.

Boiling point The temperature at which the vapour pressure of a liquid is equal to atmospheric pressure. This temperature is always the same for a particular liquid at a given pressure (for reference purposes usually taken as standard pressure). *See* also elevation of boiling point.

Boll weevil *See* Entomology.

Bomb calorimeter A sealed insulated container, used for measuring energy released during combustion of substances (*e.g.* foods and fuels).

Bond An interaction between two or more atoms or groups of atoms holding the atoms together.

Bond energy The energy involved in forming a bond, for ammonia. The bond dissociation energy is a different quantity to the bond energy. It is the energy required to break a particular bond in a compound, *e.g.*:

$$NH_3 \rightarrow NH_2 + H$$

Boracic acid *See* boric acid.

Boran Boran deficiency symptoms vary with the kind and age of the plant, the conditions of growth and the severity of the deficiency.

Borax *See* disodium tetraborate-10-water.

Borax-based test A preliminary test in qualitative inorganic analysis that can be a guide to the presence of certain metals. A borax bead is formed by heating a little borax on a loop in a platinum wire. A minute sample is introduced into the bead and the colour observed in both the oxidizing and reducing areas of the bunsen flame. The colour is also noted when the bead is cold.

Boric acid H_3BO_3 A white crystalline solid in water; in solution it is a very weak acid. Boric acid is used as a mild antiseptic eye location and was formerly used as a food preservative. It is used in glazes for enamelled objects and is a constituent of Pyres 'glass'.

Boride A compound of boron, especially one with a more electropositive element.

Borax bead colours

Metal	*Oxidizing Flame*	*Reducing Flame*
Chromium	Green H and C	Green H and C
Cobalt	Blue H and C	Blue H and C
Copper	Green H, blue C	Often opaque
Iron	Brown-red H, Yellow C	Green H and C
Manganese	Violet H and C	Colourless H and C
Nickel	Red-brown C	Grey-black C
H hot		
C cold		

Boron A hard rather brittle metalloid element of group II of the periodic table. It has the electronic structure $1s^22s^2p^1$. Boron is of low abundance (0.0003 per cent) but the natural minerals occur in very concentrated form as borax ($Na_2B_4O_7.10H_2O$) and colemanite ($Ca_2B_6O_{11}$).

Boron hydride Six hydrides of boron exist ranging from B_2H_2 to $B_{10}H_{24}$, which have complex molecular structures. The compound B_2H_6 can be prepared by reacting boron trichloride with hydrogen. A complex ion, tetrahydridoborate (III) (Borohydride, BH_4), can be formed; it has very strong reducing properties.

Boron trifluoride (Bf_3) A colourless fuming gas made by heating a mixture of boron oxide, calcium fluoride, and concentrated sulphuric acid. *See* boron trichloride.

Borosilicates Complex compounds similar to silicates, but containing BO_3 and BO_4 units in addition to the SiO_4 units. Certain crystalline borosilicate minerals are known. In addition, borosilicate glasses can be made by using boron-oxide in addition to silicon dioxide. These tend to be tougher and more heat resistant than 'normal' silicate glass.

Bosch process The reaction

$$Co + H_2O = CO_2 + H_2$$

Using water gas over a hot catalyst, used to make hydrogen for the Haber process.

Boss A small igneous intrusion with a roughly circular outline, a steep angle of contact with the surrounding and a diameter of less than 25 km. Compare stock.

Boulder A large rounded mass of rock that has been shaped by erosion and transported, for example by ice, from its source. A boulder is defined as having a diameter of over 200 mm in the UK and over 256 mm in the USA (on the Wentworth scale).

Boyle's law At a constant temperature, the pressure of a fixed mass of a gas is inversely proportional to its volume: *i.e.* pV = K, where K is a constant. The value of K depends on the temperature and on the nature of the gas. The law holds strictly only for ideal gases. Real gases follow Boyle's law at low pressure and high temperature. *See* gas laws.

Break water A barrier built into the sea to absorb the impact of waves and protect the coast.

Breccia A rock consisting of angular fragments embedded in a natural cement. It may be formed in a variety of processes, *e.g.* it can be the product of material ejected from volcanic vents, in which case it is found close to the vent, or it can be sedimentary rock occurring as thin lens-shaped beds laid down on the sides of submarine ridges in a geosyncline.

Brine A solution of water with a high concentration of salt. It occurs, for example, in salt marshes, mud flats, and some lakes with high rates of evaporation.

Bromine A deep red, moderately reactive liquid element belonging to the halogens (group VII of the periodic table). It occurs in small amounts in seawater, salt lakes, and sal deposits but is much less abundant than chlorine. Bromine reacts with most metals but generally with less vigour than chlorine.

Bronze Any of a group of copper-tin alloys usually containing 0.5–10 per cent of tin. They are generally harder, stronger in compression, and more corrosion-resistant than brass. Zinc is often added, as in gunmetal (2-4 per cent zinc) to increase strength and corrosion-resistance.

Brown aridic soils In the past covered with forest and brush, since subhumid, podzolization, though weak, is as to prevent the accumulation of lime in the profile and to render the surface oil somewhat acid.

Brown coal A type of coal with a high carbon content which is economically very important. Several varieties are found, from a soft brown coal, which appears woody and has a carbon content of 60–69 per cent to a hard brown coal, which has a carbon content of 71–77 per cent. It is found in West and East Germany and other parts of Europe.

Brown earth (brown forest soil) A widely occurring zonal soil that is associated with those areas originally covered by deciduous woodland and having moist climates with moderate rainfall.

Brownian motion The random motion of microscopic particles due to their continuous bombardment by the much smaller and invisible molecules in the surrounding liquid or gas. Particles in Brownian motion can often be seen in colloids under special conditions of illumination.

Brown podzolic soils This is a zonal group of soil of which grey infertile siliceous layer is absent or very faint, its place being taken by a friable brown or light brown A_1 horizon several inches thick.

Brown steppe soil A zonal soil of temperate grasslands occurring in drier parts than the neighbouring and similar chestnut soils. Steppe soils are a lighter brown in colour than the chestnut soils as the humus content is less and the accumulation of calcium is both more marked and more shallowly positioned in the soil profile.

Buffer A solution that resists any change in acidity or alkalinity. Buffers are important in living organisms because they guard against sudden changes in pH.

Buffer capacity of soils The higher the capacity of a soil of the greater, other factors being equal, will be its buffer capacity. This is fully recognised in practical in that the heavier the soil texture and higher its organic content, the larger must be the application of lime to effect a given change in the pH.

Bumping Violent boiling of a liquid caused when bubbles form at pressure above atmospheric pressure.

Bund An artificial quay, dam or embankment on a small ridge between fields. This word is extensively used in India.

Bunsen burner A gas burner consisting of a vertical metal tube with an adjustable air-inlet hole at the bottom. Gas is allowed into the bottom of the tube and the gas–air mixture is burnt at the top. With too little air the flame is yellow and sooty.

Burette A piece of apparatus used for the addition of variable columns of liquid in a controlled and measurable way. The burette is a long cylindrical graduated tube of uniform bore fitted with a stopcock and a small-bore exit jet, enabling a drop of liquid at a time to be added to a reaction vessel.

Burned lime It is oxide of calcium and magnesium. These have percentage of about 77 and 78 respectively of lime.

Butanoic acid (Butyric acid, C_3H_7C-OOH) A colourless liquid carboxylic acid. Esters of butanoi acid are present in butter.

Butanols (C_4H_9OH) Two alcohols derived from butane: butan-1-ol [$CH_3(CH_2)_2CH_2OH$] and butan-2-ol [$CH_3CH(OH)CH_2CH_3$]. Both are colourless volatile liquids used as solvents.

Butanone (methyl ethyl ketone, CH_3Co-C_2H_5) A colourless volatile liquid ketone. It is manufactured by the oxidation of butane and used as a solvent.

Butenedioic acid Either of two isomers. Transbutanendioic acid (fumaric acid) is a crystalline compound found in certain plants. Cisbutenedioic acid (maleic acid) is used in the manufacture of synthetic resins. It can be converted into the transisomer by heating at 120°C.

By product A substance obtained during the manufacture of a main chemical product. For example, propanone is not now manufactured from propan-1-ol but is obtained as a by-product in the manufacture of phenol.

C

Cadmium A translate metal obtained as a by-product during the extraction of zinc. It is used to protect other metals from corrosion, as a neutron absorber in nuclear reactors, and in alkali-type batteries.

Symbol Cd; m.p. 320°C; b.p. 765°C; r.d. 8.7; p.n. 48; r.a., 112.41.

Cadmium cell *See* Weston cadmium cell.

Caenozoic (Cainozoic, Cenozoic) The present geological era, beginning some 65 million years ago, and divided into two periods, the Tertiary and the Quaternary. It is characterized by the rise of modern organisms, especially mammals and flowering plants. *See* also geological time scale.

Cairn A pile of rough stones, originally of use as a memorial but commonly used today as a landmark on footpaths in mountain and moorland areas (Gaelic).

Calcareous Composed of or containing a high proportion of calcium carbonate ($CaCO_3$); the term is usually applied to rocks and soils with such a composition.

Calcification The combined process of Ca carbonate concentration in B-horizon which is influenced by grass is not mean factor is spoken as calcification. Such soils are mildly alkaline and mainly greying or yellowish-brown with white spots or blotched with spots of $CaCO_3$. The Combined process of carbonate concentration with influence of grass is known as calcification.

Calcimorphic soils A group of intrazonal soils that result from the weathering of limestones and are therefore markedly alkaline. The calcareous parent material usually prevents any development of acidity regardless of how much leaching takes place. Calcimorphic soils are usually dark, organic rich, and contain abundant soil fauna. Renzines are calcimorphic soils.

Calcite An original soil forming mineral, $CaCO_3$.

Calcium An essential mineral salt for animal and plant growth. It is present between plant cell walls and perctate, and is found in the bones and teeth of animals.

Calcium arsenate It is stomach insecticide, stronger than lead arsenate.

Calcium carbonate ($CaCO_3$) A white solid that occurs naturally in two crystalline forms calcite and aragonite.

Calcium chloride ($CaCl_2$) A white solid that occurs in a number of hydrates forms and is readily available as a by-product of the Solvay process. It is readily soluble in water and the solution known as brine, is used in refrigerating plants.

Calcium cyanamide ($CaCN_2$) A solid prepared by heating calcium dicarbide to a temperature in excess of 800°C in an atmosphere of nitrogen. It is used as fertilizer because ammonia and calcium carbonate are slowly formed when water is added:

$$CaCN_2 + 3H_2O \rightarrow CaCO_3 + 2NH_3$$

Calcium deficiency of this element first appears on the new leaves, growth of terminal buds stops and leaves become mottled. Tips and margins of leaves show drying symptoms.

Calcium dicarbide (calcium acetylide, calcium carbide, CaC_2) A colourless solid when pure. In countries where electricity is cheap, calcium dicarbide oxide is produced on a large scale by heating calcium oxide with coke at a temperature in excess of 2000°C in an electric-arc furnace. Water is then added to give ethyne (C_2H_2), an important industrial organic chemical:

$$CaC_2 + 2H_2O \rightarrow C_2H_2 + Ca(OH)_2$$

The structure of calcium dicarbide is interesting because the carbon is present as carbide ions C_2^{2-}. *See* ethyne.

Calcium exchangeable Calcium is 6 to 9 milli-equivalents. It has been also reported that all of the calcium present in organic soils, and a very large proportion of the magnesium are in an exchangeable and hence in an active condition. Approximately two-fifths of potassium is exchangeable.

Calcium-fluoride structure (fluorite structure) A form of crystal structure in which each calcium ion is surrounded by eight fluoride ions arranged at the corners of a cube and each fluoride ion is surrounded tetrahedrally by four calcium ions.

Calcium hydrogencarbonate [calcium bicarbonate, $Ca(HCO_3)_2$] A solid formed when water containing carbon dioxide dissolves calcium carbonate:

$$CaCO_3 + H_2O + CO_2 \rightarrow Ca\,(HCO_3)_2$$

Calcium hydrogencarbonate is a cause of temporary hard water. The solid is unknown at room temperature.

Calcium hydroxide [slaked lime, $Ca(OH)_2$] A white solid that dissolves sparingly in water to give the alkali known as limewater. Calcium hydroxide is manufactured by adding water to the oxide a process known as slaking, which evolves much heat. If just sufficient water is added so that the oxide turns to a fine powder, the product is slaked lime. If more water is added a thick suspension called milk of lime is formed.

Calcium metaphosphate It is a fertilizer. It has 62–63 per cent of P_2O_5. It is synthesized. It has 62 per cent phosphorus in it.

Calcium nitrate [$Ca(NO_3)_2$] A deliquescent salt that is very soluble in water. It crystallizes as the hydrate $Ca(NO_3)_2.4H_2O$. When the hydrate is heated, the anhydrous salt is first produced and this subsequently decomposes to give calcium oxide, nitrogen dioxide, and oxygen. Calcium nitrate is used as nitrogenous fertilizer.

Calcium octadecanoate [calcium stearate, $Ca(CH_3(CH_2)_{16}COO)_2$] An insoluble salt of octadecanoic acid. It is formed as 'scum' when soap, containing the soluble salt sodium octadecanoate, is mixed, with hard water containing calcium ions. *See* soap.

Calcium oxide (quicklime, CaO) A white solid formed by heating calcium in oxygen or, more widely, by the thermal decomposition of calcium carbonate. On a large scale, limestone (calcium carbonate) is heated in a tall tower called a lime kiln to a temperature of 550°C. The reversible reaction:

$$CaCO_3 = CaO + CO_2$$

Produced in a forward direction as the carbon dioxide is carried away by the upward current through the kiln. Calcium oxide is used in extractive metallurgy to produce a slag with the impurities in metal ores; it is an intermediate for the production of calcium hydroxide.

Calcium phosphate $Ca_3(PO_4)_2$ A solid that occurs naturally in the mineral apatite $[CaF_2Ca_3(PO_4)_3]$ and in rock phosphate. It is the chief constituent of animal bones and is used extensively in fertilizers.

Calcium sulphate anhydrite, $CaSO_4$ A white solid that occurs abundantly the material anhydrite and as the dehydrate ($CaSO_4.2H_2O$), known as gypsum or alabaster. When heated, gypsum loses water to form the hemihydrate ($CaSO_4.H_2O$), which is plaster of Paris. If the water is replaced, gypsum reforms and sets as a solid. Plaster of Paris is therefore used for taking casts and for setting broken limbs. Calcium sulphate is sparingly soluble in water and is a cause of permanent hardness. It is used in ceramics, paint, and in paper making.

Calcium use to alleviate soil physiology It is comparatively cheap and easily available. Moreover, its carbonates, hydroxides and oxides are mild and beneficial in their action.

Calgon A substance often added to detergents to remove unwanted chemicals that have dissolved in water and would otherwise react with soap to form a scum. Calgon consists of complicated polyphosphate molecules, which absorb dissolved calcium and magnesium ions. The metal ions become trapped within the Calgon molecules.

Caliche An impure commercial form of sodium nitrate.

Calomel *See* mercury (I) chloride.

Calorie symbol–cal A unit of energy approximately equal to 4.2 joules. It was formerly defined as the energy needed to raise the temperature of one gram of water by one degree Celsius.

Because the specific thermal capacity of water changes with temperature, this definition is not precise. The mean or thermochemical calorie (cal_{th}) is defined as 4.184 joules. The international table calorie (cal_{LT}) is defined as 4.1868 joules. Formerly the mean calorie was defined as one hundredth of the heat needed to raise one gram of water from 0X to 100°C, and the 15°C calorie as the heat needed to raise it from 14.5°C to 15.5°C.

Calorific value The energy content of a fuel, defined as the energy released in burning unit mass of the fuel.

Cannizzaro reaction The reaction of aldehydes to give alcohols and acid anions in the presence of strong bases. The aldehydes taking part in the Cannizzaro reaction lack hydrogen atoms on the carbon attached to the aldehyde group. For instance, in the presence of hot aqueous sodium hydroxide:

$$NaOH + 2C_6H_5CHO$$

$$C_6H_5CH_2OH + C_6H_5COO^-Na^+$$

The reaction is a disproportion, involving both oxidation (to acid) and reduction (to alcohol). It can also occur with methanol to give methanol and methanoate ions.

$$NaHO + 2HCHO - CH_3OH + HCOO^-Na^+$$

Capability class of soil The third and broadcast category is the 'capability classes' which places all soils under eight capability classes, are given numbers I to VIII. The risks of soil drainage of limitations in cultivation and use through progressively greater from Class I to Class VIII.

Capability sub-class of soil Second category "Sub-Class" is a grouping of capability units having similar kinds of limitation hazards are recognized, *viz.*, (1) erosion hazards, (2) wetness, (3) root zone limitations, and (4) climate.

Capability unit of soil While the first category 'capability unit' is a grouping of soils that show similar behaviour to systems of management of cultivated crops.

Capillarity In soils, the property resulting from surface tension by which water is held as a thin film around soil particles and in minute spaces (known as capillary water) may be drawn up through the soil as a result of evaporation and transpiration. The soil kept moist in this way is known as the capillary fringe. Unlike gravitational water, capillary water cannot freely drain away and remains available for plant roots unless evaporation is intense. Fine-textured clays retain a greater proportion of moisture by capillarity than do coarser soils. *See* also field capacity.

Capillary adjustments Two especially important types of interface exist in soil–liquid-solid and liquid-air. At the former interface adsorption not only of water but also of other constituents takes. Here lies the high-energy hygroscopic water. In the moisture films of the latter interface, surface tension develops an inward pressure.

Capillary formula for height of tube The height to which water rises in a capillary tube is expressed as:

$$h = \frac{2T}{rdg}$$

Here T is surface tension in dynes per centimetre, *r*, the radius of the tube in centimetres, *d*, the density of the liquid, and *g*, the gravity in dynes per centimetre.

Capillary movement and evaporation Several factors by their combined influence upon the vapour-pressure gradient account for the whole loss of water by evaporation. They are (1) the relative humidity of the atmosphere, (2) the amount of water at the soil surface and especially its rate of replacement, (3) the temperature difference of the two moisture phases, and (4) wind velocity.

Capillary and plant root extension It may be considered as how immense amount of water necessary to offset transpiration is so readily and steadily acquired by vigorously growing crops. Two phenomena seem to make adequate and continuous contact possible if the soil is in good condition of plant grown (1) The capillary adjustment of the soil water, and (2) The extension of the root system.

Capillary water Water in so called capillarity zone shows considerable range in energy values. These different energy levels are reflected in movement differences in spite of the fact that all of the water, capillary water is in a fluid state. At the higher tensions (between pF 4.2 and 4.5) capillary adjustment is very sluggish and comparatively ready movement does not occur until pressure in the neighbourhood of 2.7 (moisture equivalent) are reached.

Capillary water of soils Capillary water is capable of movement upwards, downwards or laterally, the movement always taking from thicker part of the film to the thinner part. Similarly, as the capillary water moves further away from the water table below, its rate of movement becomes less and the suction pressure needed to draw up capillary water increases. The suction force needed to withdraw a unit mass of water from a unit mass of soil and expressed in terms of the height of column of water in centimetres is called 'capillary potential.'

Carbamide *See* urea.

Carbanion An intermediate in an organic reaction in which one carbon atom is electron-rich and carries a negative charge. Carbon ions are usually formed by abstracting a hydrogen ion from a C–H bond using a base, *e.g.*, from ethanol to form CH_2CHO or from organometallic compounds in which the carbon atoms is bonded to an electro-positive metal.

Carbide A compound of carbon with a more electropositive element. The carbides of the elements are classified into ionic carbides, which contain the carbide ion C_4^-. An example is aluminium carbide, Al_4C_3. Compounds of this type react with water to give methane (they were formerly also called methanides). Another type of ionic carbon compound contains the dicarbide ion -C-C. The best-known example is calcium dicarbide, CaC_2 also known as calcium carbide, or simply carbide. Compounds of this type give acetylene with water.

Carbon The first element of group IV of the periodic table. Carbon is a universal constituent of living matter and the principle deposits of carbon compounds are derived from living sources; *i.e.* carbonates (chalk and limestone) and fossil fuels (coal, oil and gas). It also occurs in the mineral dolomite. The element forms on 0.032 per cent by mass of the Earth's crust. Minute quantities of elemental carbon also occur as graphite and diamond. The industrial demand for graphite is such that it is manufactured in large quantities using Acheson process in which coke and small amounts of asphalt or clay are raised to high temperatures. Large quantities of impure carbon are also consumed in the reductive extraction of metals. Apart from the demand for diamond as gem-stone, there is a large industrial demand for low grade small diamonds for drilling and grinding machinery. Carbon will react readily with sulphur at red heat to form CS_2 but it does not react directly with nitrogen. Cyanogen, $(CN)_2$ must be prepared by heating covalent metal cyanides such as CuCN. Carbon will also react directly with many metals at elevated temperatures to give carbides. Carbides can also be obtained by heating the metal with a hydrocarbon. There is a bewilderingly wide range of metal carbides, both salt-like with electropositive elements (CAC_2) and covalent with the metalloids (Sic), and there are also many interstitial carbides formed with metals such as Cr. Mn, Fe, Co, and Ni. Symbol C m.p. 3570°C; b.p. 4827°C; r.d. 3.51; p.n. 6; r.a.m. 12.01.

Carbonaceous Denoting sedimentary deposits formed principally from the remains of plants and other organic materials, *e.g.*, coal, peat, and petroleum.

Carbonate A salt of carbonic acid fountaining the ion CO_3.

Carbonate absorption The carbon-dioxide of atmosphere in its photosynthetic role is a direct source of most of the carbon required by plaints, but not of all. In the soil solution bicarbonates and carbonates occur in abundance, which when dissociation occurs, develop CO_3 and H_2CO_3 ions. These are absorbed by higher plants, at times a considerable amount of carbon enters the plant directly from the soil.

Carbonation The solution of carbon dioxide in a liquid under pressure, as in carbonated soft drinks.

Carbonation in soils This proceeds simultaneously with hydrolysis. The soil water is always charged with carbon dioxide taken from the atmosphere or released by the decomposition of soil organic matter and respiration of plant roots and micro-organisms. The hydroxides produced in hydrolysis react with the dissolved carbon dioxide and rapidly produce carbonates, which may either leach out or accumulate in the soil as 14 in arid regions. Carbonation is very rapid in areas of high rainfall but the leaching of carbonate in such conditions is equally expeditious.

Carbonation in weathering (carbonation-solution) A process of chemical weathering of rocks. It is produced by rainwater containing small amounts of carbon dioxide, from the atmosphere of soil, in solution (weak carbonic) when exposed to the solution the limestone, which is composed largely of calcium carbonate, is converted to calcium bicarbonate and removed in solution:

$$CaCO_3 + H_2O + CO_2 \rightarrow Ca(HCO_3)_2$$

This process is the chief erosion agent in limestone country.

Carbon-base exchange relation *See* base exchange. Some of the main stage in the carbon cycle.

Carbon-cycle of the soil Most of the energy acquired by the fauna and flora within the soil comes from the oxidation of carbon and as a result, its oxide is evolved continuously and in large amounts. The various changes that this element undergoes within and outside the soil are called the carbon cycle.

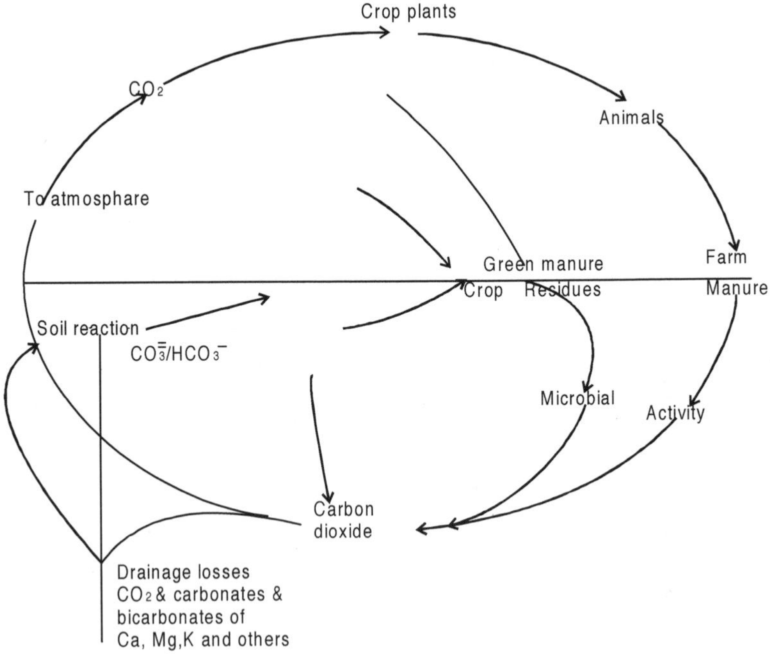

Carbon Cycle in and Outside the Soil

Carbon dating A method of dating-measuring the age of (usually archaeological) material that contain matter of living organism. Also *See* radioactive dating.

Carbon dioxide (CO_2) A colourless odourless non-flammable gas formed when carbon burns in excess oxygen. It is also produced by respiration. Carbon dioxide is present in the atmosphere (0.03 per cent by volume) and is converted in plants to carbohydrates by photosynthesis. In the laboratory it is made by the action of dilute acid on metal carbonates. Industrially, it is obtained as a by-product in certain processes, such as fermentation of the manufacture of lime. The main uses are as refrigerant (solid carbon dioxide, called dry ice) and in fire extinguishers and carbonated drinks.

Carbon dioxide is the anhydride of the weak acid carbonic acid, which is formed in water.

$$CO_2 + H_2O \rightarrow H_2CO_3$$

Carbon disulphide (CS_2) A colourless poisonous flammable liquid made from methane (natural gas) and sulphur. It is a solvent. The pure compound is virtually odourless, but CS_2 usually has a revolting smell because of the presence of other sulphur compounds.

Carbon generation in soil Important source of carbon of soil is the organic matter received by it in three ways (1) original tissue, (2) partially decomposed materials, and (3) humus.

Carbonic acid The presence of carbonic acid in soil (H_2CO_3) in the soil solution is inevitable. The ionic hydrogen thus produced by it is extremely active and tends to replace the exchangeable calcium of the colloidal complex.

Carboniferous A geological period in the Palaeozoic era extending from about 345 million years ago. It was named after the extensive deposits of coal that were formed during this period. It is divided into two parts–the upper and lower carboniferous. The lower carboniferous is composed of thick extensive deposits of carboniferous limestone formed in warm tropical seas. This was followed by the Millstone grit formed under deltaic conditions and the coal measures of the upper carboniferous which were derived from decayed and compressed swamp vegetation. The carboniferous rocks are some of the most important reserves of the iron ore, the bulk of the world's coal and also containing significant reserves of iron ore, oil and oil shale.

Carbonium ion An intermediate in an organic reaction in which one carbon atom is electron-deficient and carries by breaking a covalent bond between a carbon atom and an electronegative atom.

Carbonize (carburize) To covert an organic compound into carbon by incomplete oxidation at high temperature.

Carbon monoxide (CO) A colourless flammable toxic gas formed by the incomplete combustion of carbon. In the laboratory, it can be made by dehydrating methanoic acid with concentrated sulphuric acid.

Carbon-nitrogen ratio of the soil Carbon and nitrogen have received particular attention because of the energy relations of the former and the nutrient relations of the latter.

Carbon in soil air The soil air because of its varying amounts of carbon dioxide and oxygen presents to the roots of higher plants and to micro-organisms an atmosphere suited to almost any type of biochemical activity.

Carnallite A mineral chloride of potassium and magnesium, KCl $MgCl_2$-$6H_2O$.

Carrier-gas The gas used to carry the sample in gas chromatography.

Carriers of nitrogen Of the various organic groups, the nitrogenous materials are most probably the most completed as they carry not only carbon, oxygen and nitrogen, but such elements as sulphur, iron, phosphorus and others in lower amounts. They are compounds of high molecular weight and many are of unknown constitution.

Case hardening Processing for increasing the hardness of the surface or 'case' of the steel used to make such components as gears and crankshafts. The content of the surface layer is increased by heating in a carbon-rich environment.

Catalyst A substance that alters the rate of a chemical reaction without itself being changed chemically in the reaction.

Catchment The area from which a single river system collects its waters. The boundary of a catchment area is defined by the watershed. *See* also drainage basin, watershed.

Catecholamines A group of chemicals (amine derivatives of catechol) that act as neurohormones or neurotrans-mitters. Examples are noradrendline, adrenaline and dopamine.

Catena A sequence of soil types usually developed from similar parent material that occurs from the top to the bottom of a hillslope in an area of similar relief.

Catenation The formation of chains of atoms in molecules.

Cathode In electrolysis, the electrode that is at a negative potential with respect to the anode. In any electrical system, such as a discharge tube or electronic device, the cathode is the terminal at which electron enter the system.

Cation A positively charged ion, formed by removal of electrons from atoms or molecules. In electrolysis, cations are attracted to the negatively charged electrode (the cathode). Compare anion.

Cation exchange in soils A clay complex may be represented as x Ca r H (acidoid). The significance of Ca and H are obvious since in a humid region these cations dominate the complex. The r stands for the small amounts of cationic Mg, K, Na, etc. that are usually present in the outer ionic layer. The x indicates that the number of micaelles is indefinite.

Cations in soils While all sorts of cations may be loosely held by the adsorption power of the clay nuclei, certain ones are especially prominent. For a humid region clay are H^+ Ca^{++} first, Mg^{++} second, and K^+ and Na^+ third. For well-drained arid regions soils the order of the exchangeable ions is catt and Mg^+ first, Na^+ and K^+ next and H^+ least. The first clay is considered as having a calcium-hydrogen complex while the second is dominated by calcium and magnesium.

Caustic potash *See* potassium hydroxide.

Caustic soda *See* sodium hydroxide.

Cedar-tree laccolith An igneous instruction composed of a series of laccoliths, one above the other, in which the molten rock was intruded along several bedding places as it penetrated upwards. The top of the intrusion is domed and the lower intrusions resemble the branches of a tree. *See* also laccolith.

Cement A powdered mixture of calcium silicates and aluminates, which is made by heating limestone ($CaCO_3$) with clay, add grinding the result. When mixed with water, reaction occur with the water (hence the name hydraulic cement), and a hard solid aluminosilicate is found.

Cementite (Fe_3C) A constituent of certain cast irons and steels. The presence of cementite increases the hardness of the metal.

Centimetre-gram second system of units In this system, centimetre, gram and second are the units of length, mass and time respectively.

Centre of gravity It is the point about which the body balances. It is a fixed point in a body through which the whole weight of the body may be supposed to act vertically downwards.

Centrifugal force It is the reaction of the centripetal force, and hence it is equal in magnitude and opposite to directions to the centripetal force.

Centripetal force Developing from the outside inwards so the youngest structures are at the centre. Centripetal xylem differentiates from the outside towards the centre, and a centripetal inflorescence (*e.g.* the capitulum) is one in which the outer flowers open before those in the centre. Compare centrifugal. *See* also acropetal, basipetal.

Ceramics Useful high-melting inorganic materials. Ceramics include silicates and aluminisilicates, refractory metal oxides, and metal nitrides, borides etc. Pottery and porcelain are examples of ceramics.

Chain reaction A self-sustaining chemical reaction consisting of a series of steps, each of which initiated by the one before it. An example, is the reaction between hydrogen and chlorine.

$$Cl_2 \rightarrow 2Cl$$

$$H_2 + Cl \rightarrow HCl + H$$

$$H + Cl_2 \rightarrow HCl + Cl$$

Chain surveying A method of surveying. The instrument used is a chain which consists of a metal chain now usually 20 m (formerly 22 yards) long made up of 100 links and a set of markers, usually steel pegs which are used to mark the ends of chains as measurements are made. Chain surveys are made by surveying a series of triangles and offsets from the edge of the triangles.

Chairality *See* optical activity.

Chalk A natural form of calcium carbonate ($CaCO_3$) formed originally by marine organisms. (Blackboard chalk is calcium sulphate, $CaSO_4$). A soft white very pure limestone, consisting of about 90 per cent calcium carbonate. It is the characteristic rock of the cretaceous period in W Europe.

Chalybeate Water impregnated with iron salts. It is usually applied to mineral water or spring water.

Changes in original soil material The weathering processes not only break up rocks into small particles but also produce chemical changes in the original material. Valuable plant-food materials of vegetable matter or animal remains increases the crop producing power of soil.

Changes of lime in soil This is influenced by two factors: (1) the magnesium compounds applied undergo solution on under the influence of a variable pressures of carbon dioxide, and (2) the colloidal complex, unless already fully satisfied in respect of bases, which will into be the case with acid soils, will absorb considerable amounts of calcium and magnesium ions.

Channel A stretch of water confined between banks as a river or stream channel.

Charles' law For a given mass of gas at constant pressure, the volume increases by a constant fraction of the volumes at 0° for each Celsius degree rise in temperature. The constant fraction ($\propto$) has almost the same value for all gases–about 1/273–and Charles' law can be written in the form

$$V = V_0 (1+a_v\theta)$$

where, V is the volume at temperature 0°C. and V_0 the volume at 0°. The constant $\propto_v$ is the thermal expansivity of the gas. For an ideal gas, its value is 1/273.15. A similar relationship exists for the pressure of a gas heated at constant volume

$$P - P_0 (1 - \propto_p\theta)$$

Here, p is the pressure coefficient. For and ideal gas.

$$p = \propto_v$$

although they differ slightly for real gases. It follows from Charles' law that for a gas heated at constant pressure,

$$V/T = K$$

where, T is the thermodynamic temperature and K is a constant. Similarly, at constant volume, p/T is a constant. Charles' volume law is sometimes called Gay-Lussac' is law after its independent discoverer. *See* also absolute temperature gas laws.

Chealtion In soils, a process chemical bonding of metallic ions (notably iron, aluminium and magnesium) to organic compounds derived from the decomposition humus. The organic compounds, known as chelating agents, are extracted from the litter layer by water passing down through it. Chelating agents are richest in heath plants and conifer needles.

Chelat A metal coordination complex in which one-ligand co-ordinates at two or more points to the same metal ion. The resulting complex contains rings of atoms that include the metal atoms.

Chemical bond A link between atoms that lead to an aggregate of sufficient stability to be regarded as an independent molecular species. Chemical bonds include covalent bonds, electrovalent bonds, coordinate bonds and metallic bonds.

Chemical combination laws A group of chemical laws developed during the late 18th and early 19th centuries, which arose from the recognition of the importance of quantitative (as opposed to qualitative) study of chemical reactions. The laws are: (1) the law of conservation mass (matter) (Lavoisier, 1774), (2) the law of constant (definite) proportions (Proust, 1779), (3) the law of multiple proportions (Dalton, 1804).

These laws played a significant part in Dalton's development of his atomic theory (1808). *See* conservation of mass (law of), constant proportions (law of), multiple proportions (law of).

Chemical engineering The branch of engineering concerned with the design and maintenance of a chemical plant and its ability to withstand extremes of temperature and pressure, corrosion, and wear.

Chemical fossils Particularly resistant organic chemicals present in geological strata that are though indicate the existence of life in the period when the rocks were formed.

Chemical weathering The disintegration of rock by chemical process, most of which involve the action of water. These processes may cause the removal of cementing materials in a rock, weakening its structure so that it tends to crumble, or they may led to the formation of weaker secondary minerals, which more easily eroded. Chemical weathering is most effective kin wet environment that experience high temperatures and humidity.

Chemisorption *See* adsorption.

Chemotrophism A type of nutrition in which the source of energy of the synthesis of organic requirements is chemical. Most chemotrophic organisms are heterotrophic (*i.e.* chemoheterotrophic) and their energy sources is always and organic compound; animals, fungi, and most bacteria are chemoheterotrophs. If autotrophic (*i.e.* chemoautotrophic or chemosynthetic) the energy is obtained by oxidation of an inorganic compound.

Chemical analysis for base exchange capacity The unit is a milliequivalent and is 'one milligram of hydrogen or the amount of any other ion that will combine with or displace it.' Milliequivalents when applied to soils are usually expressed on the basis of 100 grams of dry matter.

Chemical analysis of caustic lime in soil The calcium oxide equivalent as the term implies, is a statement of the strength of the lime in one figure, namely CaO. The magnesium oxide is calculated to calcium oxide equivalent and this figure is added to the calcium oxide present already.

Chemical analysis for organic carbon The percentage organic matter in a single sample of soil when calculated from the percentage of nitrogen by the usual factor (N x 20), is an untrustworthy methods. But when the factor is applied to the average percentage of nitrogen in a large number of soils, it gives a comparatively accurate group figure especially for aerable soils.

Chemical analysis by rapid test These tests are in vogue and fulfil the need of soil analysis partially. the determination of the pH of soil acidity and the factors associated with it. As to find the status of lime also it is of great importance.

Chemical analysis of soils It is done in many aspects such as base exchange capacity, complete analysis, fertilizer need, organic carbon etc. In dealing with the fertility problems of any particular soil, it is necessary to know first of all what nutrient element or elements are deficient in the soil or are not becoming available readily enough to meet the current needs of plants, hence a chemical analysis is very much needed.

Chemical composition of silicate clays Because of the mineralogical heterogeneity of the clayey complex, settle light can be thrown on the react composition of the mineral individuals.

Chemical method of soil reclamation Soils that are already alkaline in reaction are best reclaimed by gypsum (calcium sulphate). Gypsum reacts with the soluble carbonates to form the comparatively insoluble calcium carbonate and the sulphate of sodium and potassium.

Chemical nature of humus Chemically, humus represents a mixture of carbon-hydrates, proteins, fats, resins, wax and other similar substances. These complex compounds are gradually decomposed by soil micro-organisms into simple mineral salts, carbon dioxide, water, ammonia, methane, free nitrogen and organic acids.

Chemical properties of soils The rocks from which soils are formed differ very much in their mineralogical and chemical composition. Igneous rocks are divided into acidic, intermediate and basic groups, depending their relative content of silica. Similarly, sandstones and shales differ markedly in composition.

Chemical status of black soils The black soils are usually deficient in nitrogen phosphoric acid and organic matter, but rich in potash, iron, lime alumina and calcium and magnesium carbonate.

Chemical status of red soils Profile characteristics, specially in the presence of cellular concretions of iron irrespective of their chemical composition. Detailed examination has shown that combined silica and free iron are present in laterite crusts and that the cementation action is due to the presence of free sesquioxide.

Chemical analysis of leaves This is a method to detect any elemental deficiency in plants. By such analysis it is found that in comparison with the weak plants, what are the difference that exist in the healthy or normal plants.

Chemicals used in soil testing Use is made of weak solutions carbonic, nitric, sulphuric, oxalic, citric and acetic acids to extract the available nutrients. During this process, the exchangeable calcium, magnesium, potassium and other cations are replaced and phosphorus is released. The solution thus obtained is tested, usually colorometrically, for the nutrient elements in question.

Chemical weathering In soil formation it takes place with the help of water (hydrolysis), hydration, carbonation (CO_2), and oxidation (O_2). Solutions of water changed with CO_3.

Chernozem (black earth) A black or dark-brown zonal soil that is rich in humus and contains lime.

Chestnut soil A zonal soil of mid-latitude grasslands that occurs in drier regions than chernozems.

Chimney A narrow vertical cleft in a rock face, formed where a joint plane has been expanded by weathering.

China clay *See* kaolin.

Chloramine (NH_2Cl) A colourless liquid made by reacting ammonia with sodium chlorate (l) (NaOCl). It is formed as in intermediate in the production of hydrozine. Chloramine is unstable and changes explosively into ammonium chloride and nitrogen trichloride.

Chlorate A salt of a chloric acid.

Chlorination 1. Treatment with chlorine; for instance, the use of chlorine to disinfect water. 2. *See* halogenation.

Chlorine A green reactive gaseous element belonging to the halogens (group VII of the periodic table). It occurs in seawater, salt lakes, and underground deposits of halite, NaCl. It accounts for about 0.055 per cent of the Earth's crust. Chlorine is strongly oxidizing and can only be liberated from its salts by strong oxidizing agents, such as maganese dioxide, potassium permanganate, or potassium dichromate; note that sulphuric acid is not sufficiently oxidizing to release from chlorine from chlorides.

Chlorine dioxide (ClO_2) An orange gas formed by the action of concentrated sulphuric acid on potassium chlorate. It is a powerful oxidizing agent and its explosive properties in the presence of a reducing agent were used to make one of the first matches. It is widely used in the purification of water and as a bleach in the flour and wood pulp industry.

C-horizon The soil horizon at the base or a soil profile, comprising mainly parent material in a partly weathered state. C-horizons supply the minerals to the soil. They are usually clearly distinct from the B-horizon above, to which they graduate as disintegration processes. *See* soil horizon.

Chromatography A method of analysing materials involving the separation by selective absorption of the various compounds as identifiable bands. For instance, a mixture of substances in solution is passed slowly down a long column packed with alumina. The different compounds move at different rates and separate into bands. In general chromatography involves a test material being carried by a moving phase (liquid or gas) through a stationary phase (solid or liquid).

Chromium A transition metal that occurs naturally as chromite ($FeO. Cr_2O_3$), large deposits of which are found in Zimbabwe. The ore is converted into sodium dichromate (VI) and then reduced with carbon to chromium (III) oxide and finally to metallic chromium with aluminium. Chromium is used in strong alloy steel and stainless steel and for plating articles. It is a hard silver metal that resists corrosion at normal temperatures.

Cinnabar A red mineral form of mercury (II) sulphide. It is the principal ore of mercury.

Climate The average weather conditions and variations in these conditions in both space and time over a large areas. Weather conditions over a specific length of time, usually a period of at least 30 years, are taken into consideration. The main elements of climate are temperature, atmospheric pressure wind, and humidity (including precipitation). On a large scale, the climate of a particular region is determined by:

1. Latitude and the tilt of the Earth's axis, which determine the amount of solar radiation received by the area.
2. The distribution and land and sea and proximity of ocean currents.
3. The attitude and topography of the area.

4. The location of the area in relation to the main circulation belts of the Earth. On a smaller scale other factors such as aspect may locally influence climate.

Climate and soil classification Soil genesis is a constructive process, unique distinct and interesting. Due to difference in climate in vegetation and in the original soil materials there are great differences from place to place in the intensity and sequence of soil forming forces, especially in the later stages of soil development, this difference is a qualitative one.

Climate and soil organic matter This topic is considered generally under three heads–(1) the source of supply, (2) the promotion of proper soil conditions in order that the organic matter may perform its legitimate functions, and (3) the economy on humus maintenance.

Classes I–IV of soils Soils grouped under Classes I to IV under good management including provision for drainage or soil conservation measures as necessary, are considered good for cultivated crops.

Classes V to VII of soils Class V, VI and VII are suited to the use of adapted nature plants. Thus Class V lands considered good for grasses and pasture, while class VII land can be put under specialised crops like coffee and tea which can be grown under highly intensive management involving elaborate practices for soil and water conservation. Otherwise Class VI and VII lands are to be put under natural vegetation including forestry.

Classes VIII of soil being very poor are incapable of giving satisfactory returns without major reclamation and as such are best relegated to natural vegetation and wild life.

Classification of soil material *See* classes of soil. When the unconsolidated debris lie in position above bed rock for centuries of it might sooner or later be removed to new positions by the mechanical forces of nature and accordingly two groups of material are designated as sedentary and transported.

Clastic Denoting rocks that are formed from fragments resulting from the breakdown of preexisting rocks. Sandstones, mudstones and conglomerates are the examples of such formation. This term can also be applied to recent sediments, such as the sand on a beach or deposits on the bed of river.

Clay A soil in which the particles have a diameter of less than 0.002 mm.

Clay soils Clays are formed from igneous rocks. They usually contain iron oxide and carbonaceous matter together with varying amounts of quartz, feldspar and mica. Chemically, they are hydrated aluminium silicates and may contain varying amounts of calcium, magnesium, sodium and iron.

Clay fraction In the process of decomposition, the rock minerals are gradually replaced by relatively more stable new minerals which are the products of decomposition. The clay particles formed in this manner are crystalline in nature and, as shown by X-ray and petrographic techniques, are composed of sheets of hydrates alumina and silica linked by oxygen atoms. Because of their very small size, they are highly reactive and form the seat of base exchange in soil.

Clay-humus complex The close association between the colloids of clay (mineral) and finely divided particles of humus (organic) in the soil. The clay-humus complex is chemically active in capturing exchangeable nutrients, which are then available for plants.

Claypan A compact layer within the soil consisting of clay particles that have been re-deposited after evaluation. Claypans become impervious with water-logging. *See* also hardpan. By downward movement of clay it tends to fill up pore spaces, sometimes resulting in a decidedly impervious condition called claypan or hardpan.

Clay's plasticity, cohesion and flocculation These properties of clays are important from the practical as well as technical point of view.

Clays of salt acid like nature Clays apart from their mineralogical character, they are made up of a complex negative radical, the acidoid or micelle, and a miscellany of adsorbed cation, of which in a humid soil, those off calcium and hydrogen are by far the most numerous.

Cleavage The tendency of some rocks to split into thin sheets or slabs along parallel plates.

Climatology The study of climate including study of climatic phenomena, their causes and their influence on the natural environment.

Cloud seeding *See* crop science.

Clue to nutrient requirement It is also sometimes possible to obtain a clue to the nutrient deficiencies of soil with the help of deficiency. Correct diagnosis of deficiency symptoms needs extensive experience. Furthermore, such symptoms in plants appear long after the actual occurrence of nutrient deficiency in soil. Therefore, such soil deficiencies must be diagnosed and remedied much earlier by other means.

Coagulating capacity of cations in soils The various coagulate in the order the H and Ca, Mg, K, Na. As the colloidal complexes of the humid regions soils are usually dominated by calcium and hydrogen. Therefore, such soils gradually tend to assume a coagulated condition. This is the association of particles (*e.g.* in colloids) into clusters.

Coal A carbonaceous rock. It is formed from the compaction and heating under pressure of large quantities of partially decomposed vegetable matter such as peat that accumulated in vast coastal swamps. There are two major groups: humic coals, which are derived from wood and include bituminous coals, lignite and anthracite; and sapropelic coals, which are derived from spores, algae and fine particles of plant material, and include cannel coal and boghead coal.

Coalescence In meteorology the process whereby where droplets of water in a cloud are enlarged by combination with other droplets. The size of the droplets will depend on several factors including the liquid content of the cloud, the amount of variation in sizes of droplets and the length of time the droplets live.

Cobalt A malleable metal that is used in the production of alloys such as stainless steel, in electroplating, and in the manufacture of blue pigments for the glasses, enamel and pottery industries. It naturally occurs as Cobaltite and is also found in copper, nickel, iron, silver and lead ores. The chief producers are Zaira, Zimbavwe and Canada.

Cobalt (II) oxide (cobaltous oxide, CoO) A green powder prepared by the action of heat on cobalt (II hydroxide in the absence of air. Alternatively it may be prepared by the thermal decomposition of cobalt (II) salts. It is stable in air upto temperatures of around 600°C, after which it absorbs oxygen to form tricobalt tetroxide. Cobalt (II) oxide can be reduced to cobalt by heating in a steam of carbon monoxide or hydrogen. It is used in the pottery industry and in the production of vitreous enamels.

Coefficient of friction It is the ratio of the limiting friction F to normal reaction R, for any two surfaces *i.e.* $u = \frac{F}{H}$.

Cohesion A second characteristic closely related to plasticity is cohesion. As water of the clay gel is reduced, a shrinkage occurs and various structural forms develop depending on the nature of the soil involved. The shrinkage, which splits the soil mass into aggregates, is accompanied by a cohesion of the particles which makes structural development not only possible but also more or less permanent.

Cohesion in organic soils While decayed organic matter, in large degree colloidal and possess high adsorptive and catalytic powers, its cohesion and plasticity are low.

Cohesion in soils Soils, because of alternate imbibition and shrinkage, accompanied and cohesion and influenced by soluble salts, develop characteristic profile structures. These are important in field classification and description. Cohesion is the tendency of particles to stick to each other. Plastic soils are cohesive.

Coke A fuel consisting chiefly of carbon, obtained from heating coal in an oven from which air is excluded, thereby removing some of the chemical substances as gases. Coke is an essential ingredient in the manufacture of pig iron and is also uses in production of steel.

Colligative properties A group of properties of solutions that depends on the number of particles present, rather than the nature of the particles. Colligative properties include:

1. The lowering of vapour pressure.
2. The elevation of boiling point.
3. The lowering of freezing point.
4. Osmotic pressure.

Colligative properties are all based upon empirical observation. The explanation of these closely related phenomena depends on intermolecular forces and the kinetic behaviour of the particles, which is qualitatively similar to those use in deriving the kinetic theory of gases.

Colour of soil The colour of the soil often indicates other important soil conditions. The surface soil is usually of a darker colour than the sub-soil. The soils vary in colour black, dark brown, red, reddish brown, yellow, bluish green, greenish grey, and grey to white. Generally speaking dark brown to black colour of soil denotes a high content of humus and the associated good soil structure and adequacy of calcium and nitrogen. Such soils are usually regarded very productive. The colour of soils in depressions and low spots, which usually remain wet in humid regions, is also black. The red or brown colour of laterites is closely associated with good aeration and drainage. In well drained soil, ferric compounds of iron are commonly formed, giving it a red or brown colour. On the other hand, in the imperfectly drained soil, iron oxides are formed, giving the soil a yellow colour. In humid regions, red or yellow soils can be turned brown by incorporation larger quantities of organic matter. The sandy deserts have the natural grey colour of their parent rocks. The reduction compounds of iron give the soils grey or greyish blue colour. Their presence is injurious to root growth. Though leaching organic matter iron compounds in wet areas may also produce a grey colour in soils. Accumulation of alkali salts makes soils white or black, depending on the salts. Such soils are the least fertile.

Colour of rock material Rock weathering iron and its discolouration due to oxidation is an important indication of chemical decay. The yellow colouration of limonite is indication of this kind of decay. Reds and yellow in all gradation and shades may be expected in soil material where vigorous oxidation and hydration of iron compounds has taken place. The presence of iron, as already is a very important factor in rock weathering, and the discolouration due to its oxidation is an unfailing indication of chemical decay. It seems that many mineral soils inherit their colours, at least in part and with modifications.

Humus has distinct colour effects, especially in surface soil layers and with modification, because weathering as a profile develops, may alert in some degree and shades and tints to the parent material.

Colorimetric analysis Quantitative analysis in which the concentration of a coloured solute is measured by the intensity of the colour. The test solution can be compared against standard solutions.

Colloid In soil science, a minute division of the soil, either mineral (clay) or organic (made up of finely divided humus). Without a colloidal fraction, soils are physically dry and chemically inert. The vast surface area of colloids has electrical attraction for mineral ions released during weathering or contained in soil water. As a result colloids hold a vital reservoir of exchangeable plant nutrients. Their water retention allows plant roots access to moisture that would otherwise drain away. Colloids swell on wetting, thus giving plasticity to the soil structure. *See* also clay-humus complex. *See* soil.

Colloid-biological concept of soil The soil is visualized as possessing a framework of mineral matter in part non-colloidal and inert; in part colloidal and extremely active. In mineral soils clayey materials make of much of this latter fraction. Mixed with the viscous, gelatinous mineral, and partially adsorbed thereby a certain amount of humus; colloidal and unstable.

Colloidal clay Soil particles of 0.002 mm or less diameter have been defined as clay. The smallest particles in the clay fraction (1 micron, *i.e.*, 0.001 mm or less in diameter) are termed colloidal clay.

Colloidal state of matter Many problems especially those relating to the fenesis, profile characteristics, and the normal functioning of the natural body of colloid are all concerned with the subject of soil. Colloidal refers to a 'two phase system in which one material in a very fine state of aggregate, approaching but not attaining a molecular subdivision, is heterogeneously dispersed through a second.

Colluvial Formed by land slides and avalanches on steep hill slopes caused by gravity; found largely at the base of mounts.

Colluvial debris It is made up of the fragments of rocks detached from the heights above and carried down the slopes by gravity. Frost action has much to do with the development of such debris.

Colluvial soil material This type of soil is made up of the fragments of rocks detached from the heights above and carried down the slopes by gravity. Soil material developed from colluvial accumulation is usually coarse and stony, as physical rather than chemical weathering has been dominated. Such deposits are not of much importance agriculturally.

Colluvium A deposit consisting of a wide variety of particle sizes ranging from rock fragments to fine grained soil which accumulates at the base of slopes under gravity.

Comber test A sample of air-dry soil held in a test tube is treated with about double of its volume of an alcoholic solution of potassium sulfocynate (KSCN). The contents of the tube are then well shaken and allowed to stand for an adequate time. The intensity of the red colour that develops in the supernatant liquid, if the soil is acid, an indication of need for lime.

Complete analysis of soil It consists of finding out all the constituents of representative sample of surface soil, expressed in percentage.

Complex A type of compound in which molecules or ions form coordinate bonds with a metal atoms or ion.

Complex forms of nutrients in soils Some of these forms are very complex and less active such as organic combinations, proteins, amino acids, etc. for nitrogen. For phosphorus, they are apatite, and secondary Ca, Fe, Al phosphates, potassium in original mineral 1 forms such as feldspar and mica etc. and others

Complex intermediates	Decomposition	Resins	Amides
Products of decay	Compounds	Waxes, oils and fats	Alcohols
Resistant compounds	Amino cells		

Component materials of soil horizons The soil horizons are made up of solid, liquid and gaseous materials. The solid portion consists of inorganic and organic substances. The inorganic portion is composed of the substances formed by the chemical and physical weathering of parent rocks, while the organic portion consists of living plants and animals and their residues.

Composition of arid soils Roughly speaking, these have 3.25 per cent organic matter, 0.12 per cent nitrogen (N_2), 0.15 per cent phosphoric acid (P_2O_5), 2.40 per cent potash (K_2O), 1.50 per cent (CAO), 1.00 per cent magnesia (MgO), 0.20 per cent sulphur trioxide (SO_3).

Composition of clay/granite When compare a chemical analysis of clay with that of the unweathered rock mass beneath, it is true that granite rock analysed aid of the same composition as that from which the clay soil material is derived.

Composition of humid soils These soils have approximately this composition. Organic matter 4 per cent, nitrogen (N_2) 0.15 per cent, phosphoric acid (P_2O_5) 0.10 per cent, potash (K_2O) 2.0 per cent, lime (CaO) 0.60 per cent, magnesia (MgO) 0.50 per cent and sulphur trioxide (SO_3) 0.10 per cent.

Composition of soils air Like atmospheric air, the soil air is composed largely of nitrogen and oxygen, but differs from the former in containing more moisture, more carbon dioxide and a little less oxygen. The actual content of carbon dioxide varies greatly during the crop season, depending on the rate of crop growth and sub-soil biological activity. Besides free air, the soil also contains air held by colloids or dissolved in soil solution.

Computation of soil data In order to have a constant and comparable basis for calculations, almost all soil data are computed on dry soil. Consequently, unless something is said to the contrary, this could always be assumed.

Composition of velocities or forces When a body possesses a number of velocities simultaneously or a number of forces are acting on a body simultaneously, then their total effect can be represented by a single velocity or force respectively known as resultant velocity or resultant force. The process of finding the resultant is called composition of velocities or forces.

Compressed air sprayer An sprayer which can be operated by one man. First the air pressure is pumped into the chamber and then spray nozzle is operated till the pressure becomes too low for good spraying.

Compression The squeezing action that produces folds in the rocks of the Earth's crust; the tightness of the folds is an indication of the strength of the squeezing.

Compound A chemical combination of atoms of different elements to form a substance in which the ratio of combining atoms remain fixed and is specific to that substance.

Compounds of lime Three compounds of lime are in common use: oxides, hydroxides and carbonates. All these have the advantage of not leaving any residue in the soil.

Concentrated Denoting a solution in which the amount of solute in the solvent is relatively high. The term is always relative; for example, whereas concentrated sulphuric acid may contain 96 per cent H_2SO_4, concentrated potassium chlorate may contain as little as 10 per cent $KClO_3$. Compare dilute.

Concentration The amount of substance per unit volume in a solution. The unit is the mol dm^{-3}. Mass concentration is mass of solute per unit volume. Molar concentration is amount of substance per kilogram of solute.

Concretion A concentration of a particular mineral within a sedimentary rock, usually as an irregularly shaped nodule. Concretions are formed at or near the surface while the deposit is being transformed from loose sediment into rock, and as a concentration during the deposition of the sediment. Manganese oxide nodules are forming today in deep sea environment.

Condensation A type of chemical reaction in which two molecules join together to form a larger molecule, with the associated production of a small molecule such as water (H_2O). The conversion of a gas or vapour into a liquid or solid by cooling.

Condensation reaction A reaction in which addition of two molecules occurs followed by elimination of a smaller molecule, usually water condensation reactions (addition-elimination reactions) are characteristic of aldehydes and ketones reacting with a range of nucleophiles.

Conglomerate A clastic sedimentary rock consisting of more or less rounded pebbles held together by a cement, usually sandstone or limestone.

Conjugate acid *See* acid.

Conjugate base *See* base.

Conjugated Describing compounds that have alterating double and single bonds in their structure. But-1,3-ene (H_2C-$CHCHCH_3$) is a conjugated compound. In such compounds, there is delocalization of the electrons in the double bonds.

Conservation of soil and water Productive land is the source of human sustenance and security. Economic stability and wise use of land are inseparable. The future of the country and its teeming millions depends to a large extent on the conservation of the land resources and the maintenance of a permanently productive agriculture through proper use and treatment of land.

Conservation of momentum If no external force is acting on a system of bodies and they are acting and reacting on each other, then the sum of their moments is constant.

Conservation of energy, law of, Enunciated by Helmboltz in 1847, this law states that in all processes occurring in an isolated system the energy of the system remains constant. The law does of course permit energy to be converted from one form to another.

Conservation of mass, law of, Formulated by Lavoisier in 1774, this law states that matter cannot be created or destroyed. Thus in a chemical reaction the total mass of the products equals the total mass of the reactants (the term 'mass') must include any solids, liquids, and gases-including air–that participate).

Constant composition, law of: *See* constant proportions.

Constant proportions, Formulated by Pronst in 1779 after the analysis of a large number of compounds the principal that the proportion of each element in a compound is fixed or constant. It follows that the composition of a pure chemical compound is independent of the method of preparation. It is also called the law of definite proportions and the law of constant composition.

Contact of soil solution The contact of plant roots with soil solution affects apparently the root exudes, which accentuate the microbial activity in the immediate neighbourhood of the absorbing surfaces.

Contact placement According to this method, seed and fertilizers are sown together. The aim in view is to bring the seed and fertilizer in contact with each other. This is very practicable in wheat, barley, sorghum, pennisetum, and cotton and others.

Constant-boiling mixture A general observation for most liquids is that the vapour phase above a liquid is richer in the more volatile component (deviation from Raoult's law). Consequently most liquid mixtures show a regular increase in the boiling point as the liquid is progressively distilled.

Contour A line drawn on a map joining all places with the same height above or below mean sea level. The relief of an area can be shown on a map by drawing a series of contours representing different heights.

Control of alkali Although, physical and chemical means of eradication are practised, the control lies fundamentally in careful irrigation and a reduction of any surface evaporation that tends to cause an upward movement of salts. Irrigation practice thus not only calls for the usual moisture regulations but some corollary features as well.

Control of water run-off Strip cropping is essentially another form of rotation. Its importance in controlling run-off erosion and thereby maintaining the fertility of the soil is now universally recognized. Strip cropping in effect employs several good farming practices, such as crop rotation, contour cultivation, proper tillage, stubble mulching, cover cropping etc. Strip cropping is of the following different forms.

1. Contour strip cropping,
2. Field strip cropping,
3. Wind strip cropping,
4. Permanent or temporary buffer strip cropping.

Conversion factors for lime calculations Since the oxide form of expression is so commonly used, this type of surety is used as the conventional method. The calcium oxide equivalent, is a statement of the strength of the lime in one figure, namely CaO. The magnesium oxide is calculated to calcium oxide equivalent and this figure is added to the calcium oxide already present. This is generally done by using conversion factor.

Copper A reddish-brown metal. It is widely distributed and found commonly in both the pure native state and as an ore in combination with other elements. It is a comparatively soft but very tough metal and has widespread use in industry. As it is an excellent conductor of electricity and is also malleable the most important use of copper is in the electrical industry.

Core sample A sample of rock, soil, or ice obtained by drilling into such material with a hollow bore and withdrawing the sample intact. The technique is used to obtain samples of material from the ocean floor for oceanographic studies.

Correlation A statistical technique to determine the extent to which a relationship exists between varieties of a crop. This can be calculated by measuring the correlation coefficient *i.e.* the degree of association between two paired variables.

Corrosion The wearing away of a solid rock surface by material that is being transported (*i.e.* by water, ice, wind, or waves) of that is moving under gravity. The most common form of corrosion is that produced by a river and its load. The wearing away of rocks by chemical action covers a wide variety of processes, including solution, hydration and oxidation. *See* chemical weathering. Reaction of a metal with an acid, oxygen, or other compound with destruction of the surface of the metal. Rusting is a common form of corrosion.

Corundum (emery, Al_2O_3) A naturally occurring form of aluminium oxide that sometimes contains small amounts of iron and silicon (IV) oxide. Ruby and sapphire are impure crystalline forms. It is used in various polishes, abrasives, and grinding wheels.

Couple To equal and unlike parallel forces acting at different points on a rigid body constitute a couple. Moment of the couple about any axis of rotation is equal to the product of either force and the perpendicular distance between the two forces. Moment of the couple is P x AS, where AB is the arm of the couple.

Coupling A chemical reaction in which two groups or molecules joint together. An example is the formation of azo compounds.

Cresols *See* methylphenols.

Critical moisture When there is scarcity of water in the soil, plants do make some adjustment to the diminished intake, but soon the tension is such that the absorption will barely meet the transpiration losses. The plant is now at a precarious state. At this point one or two things may occur-depending on the nature of plants–Either the plant is forced into a resting state or it wilts. If water is not applied, death results in the latter case.

Critical point The condition of temperature and pressure under which a liquid being heated in a closed vessel becomes indistinguishable from the gas or vapour phase. At temperatures below the critical temperatures (T_C) the substance can be liquefied by applying pressure; at temperatures above T_C this is not possible.

Critical pressure The lowest pressure needed to bring about liquefaction of a gas at its critical temperature.

Critical temperature The temperature below which a gas can be liquefied by applying pressure and above which no amount of pressure is sufficient to bring about liquefaction. Some gases have critical temperatures above room temperature (*e.g.* carbon dioxide 31.1°C and chlorine 144°C) and have been known in the liquid state for many years.

Critical volume The volume of one mole of a substance at its critical point.

Crumb structure A physical property of soil in which particles join together in loosely held spheroidal masses, which resemble bread crumbs in appearance. The crumb structure provides the best conditions for plant growth; it allows air and water to move freely between the aggregates, retains nutrients, and is more easily penetrated by plant roots.

Crust The outermost shell of the Earth. It lies above the Mohorovicic discontinuity and varies in thickness from about 6 km below the ocean to about 60 km under the high mountain ranges of continents. It comprises two dominant groups of rocks; the discontinuous sial (continental crust) and sima (oceanic crust). *See* also sial, sima.

Cryolite *See* sodium hexafluoroaluminate.

Crystal A solid substance that has a definite geometric shape. A crystal has fixed angles between its faces, which have distinct edges. The crystal will sparkle if the faces are able to reflect light. The constant angels are caused by the regular arrangements of particles (atoms, ions or molecules) in the crystal.

Crystalline Denoting a substance that forms crystals. Crystalline substances have a regular internal arrangement of atoms, even though they may not exist as geometrically regular crystals. For instance, lead (and other metals) are crystalline. Such substances are composed of accumulations of tiny crystals. Compare amorphous.

Crystalline rock A rock formed by the process of crystallization; *i.e.* an igneous or a metamorphic rock. In an igneous rock, the shape and size of the crystals depends upon the rate of cooling of the magna and the order in which the minerals crystallized.

Crystallite A small crystal that has the potential to grow larger. It is often used in mineralogy to describe specimens that contain accumulations of many minute crystals of unknown chemical composition and crystal structure.

Crystallization The process of forming crystals. When a substance cools from the gaseous or liquid state to the solid state, crystallization occurs. Crystals will also form from a solution saturated with a solute.

Crystal structure The particular repeating arrangement of atoms (molecules or ions) in a crystal. 'Structure' refers to the internal arrangement of particles, not the external appearance.

Cubic Denoting a crystal in which the unit cell is a cube. In a simple cubic crystal the particles are arranged at the corners of a cube. *See* also body-centred cubic crystal, face-centred cubic crystal, and crystal system.

Curie symbol Ci A unit of radioactivity, equivalent to the amount of a given radioactive substance that produces 3.7×10^{10} disintegrations per second.

Cyanamide process An industrial process for fixing nitrogen by heating calcium dicarbide in air. $CaC_2 + N_2 \rightarrow CaCN_2$ The product, calcium cyanamide, hydrolyess to ammonia and can be used as a fertilizer. The process is expensive due to the high cost of calcium dicarbide. *See* also nitrogen fixation.

Cyanide 1. A salt of hydrogen cyanide, containing the cyanide ion (CN^-). 2. *See* nitrile.

Cyanide process A technique used for the extraction of gold from its ores. After crushing the gold ore to a fine powder, it is agitated with a very dilute solution of potassium cyanide. The gold is dissolved by the cyanide to form potassium aurocyanide [$KAu(CN)_2$]. This complex is reduced with zinc, filtered, and melted to obtain the pure gold metal.

Cyanogen (CN_2) A toxic flammable gas prepared by heating mercury cyanide. *See* also pseudohalogen.

Cycle A completed series of events that follows or is followed by another series of similar events occurring in the same order.

Cycle of erosion A model of development of landform based on the idea that they pass through a series of changes from youth to maturity to old age. The cycle begins with a recently uplifted landmass of little relief. This is eroded by fast flowing streams with many waterfalls and rapids to form a table land dissected by V-shaped valley. This is the youthful stage. This follows the other stages.

Cyclic Compound A compound containing a ring of atoms. If the atoms forming the ring are all the same the compound is homocyclic; if different atoms are involved, it is heterocyclic.

Cyclization A reaction in which a straight-chain compound is converted into a cyclic compound.

Cyclopentadiene A cyclic hydrocarbon made by cracking petroleum. The molecules have a five-membered ring containing two carbon-carbon double bonds and one CH_2 group. It forms the negative cyclopentadienyl ion C_5H_5, present k in sandwich compounds, such as ferrocene.

D

Dalton's atomic theory A theory explaining the formation of compounds by elements first published by John Dalton in 1803. It was the first modern attempt to describe chemical behaviour in terms of atoms. The theory was based on certain postulates:

1. All elements are composed of small particles, which Dalton called atoms.
2. All atoms of the same element are identical.
3. Atoms can neither be created nor destroyed.
4. Atoms combine to form 'compound atoms' (*i.e.* molecules) in simple ratios. Dalton also suggested symbols for the elements. The theory was used to explain the law of conservation of mass and the laws of chemical combination.

Dalton's law (of partial pressures) The pressure of a mixture of gases is the sum of the partial pressures of each individual constituent (This partial pressure of a gas in a mixture is the pressure that it would exert if it alone were present in the container). Dalton's law is strictly true only for ideal gases.

Daniell cell A type of primary cell consisting of two electrodes in different electrolytes separated by a porous partition. The positive electrode is copper immersed in copper (II) sulphate solution.

The negative electrode is zinc-mercury amalgam in either dilute sulphuric acid or zinc sulphate solution.

Dark rays *See* heat in soil.

Dating techniques Methods used in determining the age of rocks, fossils or archaeological remains. There are two main methods. Relative dating assesses the age of a specimen in comparison to other specimens. Absolute dating involves assessing the actual age of a specimen by using some reliable measure to time. *See* also dendrochronology, radioactive dating.

Datum level (datum plane, datum line) A horizontal line of point (zero reference) from which heights and depths are calculated. It is usually mean seal level. *See* also ordinance Datum.

Debris The loose assorted fragments of rock material sand, and clay produced by weathering and rock disintegration, consisting of a wide range or particle shapes and sizes.

Deca Symbol Da A prefix denoting 10. For example, 1 Decametre (dam)-10 meters (m).

Decahydrate A crystalline solid containing ten molecules of water of crystallization per molecule of compound.

Deci Symbol D A prefix denoting 10^{-1}. For example, 1 Decimetre (Dm)–10^{-1} metre (m).

Deceleration Same as retardation.

Decomposition and formation of soil By this process, soils are formed under the influence of chemical acrions.

Decrepitation Heating a crystalline solid so that it emits a crackling noise, usually as a result of loss of water of crystallization.

Deep weathering The process by which rocks that lies at some depth in the ground are broken down through chemical weathering to produce a thick layer of regolith (weathered material).

Defect An irregularity in the ordered arrangement of particles in a crystal lattice. There are two main types of defect in crystals.

Definite proportions *See* constant proportion, law of soil.

Deflation (geomorphology) The transportation of fine particles of sand and dust by wind action. The finest material may be carried considerable distances by dust storms to be finally deposited as loses, where deflation is concentrated in one area a deflation hollow may form.

Defloculated condition of soil The soils differ greatly in texture and consistency, ranging from sands through loams and silts to heavy clays that are ill-drained and sometimes charged with injurious accumulations of sodium salts, producing a sterile, defloculated condition called usar.

Deforestation The removal of the tree cover of an area by felling or burning. The process is usually deliberate in order to make the land available for other uses (*e.g.* agriculture). Se also reafforestation.

Deformation A change in layer of rock after it has been deposited including folding, and tilting, caused by earth movements and igneous activity.

Degenerate Describing different quantum states that have the same energy. For instance, the five orbitals in a transition-metal atom all have the same energy but different values of the magnetic quantum number m. Differences in energy occur if a magnetic field is applied or if, the arrangement of ligands around the atom is not symmetrical. The degeneracy is then said to be 'lifted'.

Degradation A chemical involving the decomposition of a molecule into simpler molecules, usually in stages. The Hoffman degradation of amides is an example.

Degrees of freedom The independent ways in which particles can take up energy. In a monatomic gas, such as helium or argon, the atoms have three transitional degrees of freedom (corresponding to motion in three mutually perpendicular directions). The mean energy per atom for each degree of freedom is KT/2, where K is the Boltzmann constant; the mean energy per atom is thus 3kt/2.

Deliquescent Describing a solid compound that absorbs water from the atmosphere, eventually forming a solution. *See* also hygroscopic.

Deltas Much of the sediment carried by streams is not deposited in the flood plain but is discharged into the body of water to which the stream is tributary. Unless there is sufficient current and wave action, the suspended material, in part, accumulates, forming delta.

Delta deposit of sediment of soil Much of the sediment carried by streams is not deposited in the flood plain but is discharged into the body of to which the stream is tributary. Unless there is sufficient current and wave action, the suspended material, in part accumulates, forming the delta.

Dendritic growth Growth of crystals in a branching habit.

Denitrification The chemical reduction of nitrate by soil bacteria. The process is important in terms of soil fertility since the products of denitrification (*e.g.* nitrites and ammonia) cannot be used by plants as a nitrogen source.

Density symbol–d The mass per unit volume of a given substance. The unit are g dm^{-3}, etc. *See* also relative density. The mass of a substance per unit volume. It is expressed in kg cm^{-3}.

Denudation The wearing away of the land surface; it includes the processes of weathering, mass movement, erosion, and transportation (The term is sometimes used synonymously with erosion but this technically excludes weathering).

Depolarizer A substance used in a voltaic cell to prevent polarization. Hydrogen bubbles forming on the electrode can be removed by an oxidizing agent, such as manganese (IV) oxide.

Deposition The laying down of material that has been weathered, eroded and transported by natural processes, such as water, wind and ice.

Derived units The units which are not independent and are derived from fundamental units are such units, *e.g.*, the unit of area contains the unit of length twice, similarly the units of volume, velocity, force acceleration etc. are derived units.

Desert soils These belong to the important zonal group of soils. These have sometimes grey colour and called as grey desert or Sierozems.

Desiccation A process of drying up, by which the moisture is removed from a material. The term is most often applied to bodies of water, such as lakes and streams, or to deposits that have been water-logged, such as muds and clays.

Desiccation cracks (mud cracks, sun cracks) Cracks that develop in fine-grained material, such as mud and clay, as it dries out through evaporation. These form polygonal pattern, particularly when the material is homogeneous.

Dessacior A laboratory apparatus for drying solids or for keeping solids free of moisture. It is a container in which is kept a hygroscopic material (*e.g.* calcium oxide, or silica gel).

Destructive distillation The process of heating an organic substance in the absence of air to produce volatile products, which are subsequently condensed. The destructive distillation of coal was the process for manufacturing coal gas and coal tar. Formerly, methanol was made by the destructive distillation of wood.

Detailed reconnaissance This survey consists in first carrying out reconnaissance survey of selected areas and mapping soil series or association in the scale of 1=1 mile. Subsequently, 10–15 p.c. of each soil area containing already fixed series or associations are covered by detailed survey in fixing up variations in soil types and phases within the series.

Deterant to soil bacteria These bacteria also cease activity when pH value of soil falls below 5.0, but become active again if the soil is limed to raise it pH value to about 6.0. The large supplies of carbon in the raw organic matter favour an enormous multiplication of bacteria, which use up all the available nitrogen formed by the nitrifying bacteria and thus cause a temporary disappearance of nitrates from the soil.

Detergents A group of substances that improve the cleansing action of solvents, particularly water. The majority of detergents, including soap, have the same basic structure. Their molecules have a hydrocarbon chain (tail) that does not attract water molecules. The tail is said to be hydrophobic (water heating). Attached to this tail is a small group (head) that readily ionizes and attracts water molecules. It is said to be hydrophilic (water loving). Detergents reduce the surface tension of water and thus improve its wetting power.

Determination of colloidal matter through hygroscopic coefficient Separate from the soil a sample of colloidal matter by means of a high powered centrifuge. When air dried, the adsorptive power of this colloidal material for water vapour is determined by exposure over a 3.3 per cent water solution of sulphuric acid. At the same time, the adsorptive capacity for water vapour is determined on an air dried sample of whole soil.

Determination of hygroscopic coefficient The method of determining hygroscopicity of a soil at any given vapour pressure is dry soil in thin layer is exposed to an atmosphere of definite humidity under conditions of constant temperature. When equilibrium is reached, the moisture content of the soil is determined.

Determination of soil colloidal matter A sample of colloidal matter of soil is obtained by after sufficient stirring of a sample of colloidal matter by means of a high-powered centrifuge. When air-dried, the adsorptive capacity of this colloidal material for water vapour is determined by exposure over 3.3 per cent water solution of sulphuric acid.

Determination of soil pH The more accurate method of determining soil pH is by means of a potentiometer, an electrometric method in which the H-ion concentration of the soil solution is balanced against a hydrogen electrode or and electrode that functions in a similar way.

Determining the critical moisture Since the wilting of a plant is determined in large degree by the tenacity with which the soil water is held, it is to be expected that this phenomenon will occur at about the same pF for every normal soil.

Determining wilting coefficient The formula used for calculating wilting coefficient are:

$$\text{Wilting Coeff.} = \frac{\text{Hygro. coeff.}}{0.68}$$

$$\text{or Wilting Coeff.} = \frac{\text{Moisture equivalent}}{1.84}$$

Development of granulation Granules develop from two distinct sources–(1) by accretion of build-up from smaller units, and (2) by the disintegration of larger and less stable aggregates. In cultivated soils, the former origin probably is the more common.

Development of hydroxide clays Hydrolysis as well as oxidation become intense and silicate minerals succumb. In many parts of tropics, wet and dry season alternate, which no doubt, greatly intensifies chemical activities especially those of the organic matter.

Development of soil Soil formation is a very slow but a ceaseless process Climate, particularly rainfall, acting over a long period of time, is the most important factor responsible for successive stages in soil development. Under warm and humid conditions, rocks weather relatively quickly and soils get leached of basic materials like lime. On the other hand, in arid or semi-arid conditions, soils usually retain considerable amounts of basic compounds such as carbonate and sulphate of calcium, magnesium and sodium, and chloride and bicarbonate of sodium.

Diagonal relationship There is a general trend in the periodic table for electro-negativity to increase from left to right along a period and decrease down a group.

Dextrose Same as glucose.

Diamond An allotrope of carbon. It is the hardest naturally occurring substance and is used for jewellery and, industrially, for drilling equipment.

Diastrophism The movements causing the large-scale deformation of the Earth's crust that produces the continents, oceans, mountains, etc.

Diatom ooze A siliceous ooze formed from the minute particles of silica that remain when microscopic diatom algae, which are found the colder surface water of the ocean, die and decompose. It is found mainly on the ocean floor beneath these colder waters, for examples, in a broad belt beneath the Southern Ocean.

Diatomic Describing a molecule that consists of two atoms. Hydrogen (H_2), oxygen (O_2), nitrogen (N_2) and the halogens are examples of diatomic elements.

Differential accumulation of minerals and salts In the course of weathering of rocks, the readily soluble products are carried away in drainage waters, whereas comparatively insoluble products accumulate to form soil. This is particularly so in humid climates, and as a consequence, silica, alumina and iron oxide, as such or in silicate combinations, form the major portions of soils in heavy rainfall areas. Silicon in silicate form also dissolves, leaving laterites, which consists chiefly of aluminium iron oxides. In arid regions, leaching plays an insignificant part and, therefore, the chemical composition of soils is very similar to that of the rocks from which they originate.

Differential denudation A process that occurs in rocks that have varying resistance to the action of weathering and erosion. More resistant rocks, such as granite, will be weathered and eroded less quickly than comparatively soft rocks, such as clay and mudstone, producing landforms and landscapes strongly determined by rock type.

Differential heating and cooling of rocks Due to slow conduction of heat, the outer surface of rock mass often maintains a markedly different temperature than the inner and more protected portion. This differential heating and cooling lead to set up lateral stresses which in time, may cause the surface layers to peel off from the parents mass. This phenomenon is also called exfoliation.

Differential soil nutrients It may be observed that the plant nutrients form a very small part of the soil and that a soil may be rich in one ingredient and poor in another. Though plants obtain most of their nourishment, and particularly nitrogen, from the surface soil, yet they secure a certain quantity from the lower layers also. Plant roots ordinarily penetrate to a depth of three or four feet but under arid or semi-arid conditions and in well drained soils, they penetrate down to eight or ten feet.

Dilute Denoting a solution in which the amount of solute is low relative to that of the solvent. The term is always relative to that of the solvent.

Dimer A compound (or molecule) formed by combination or association of two molecules. For instance, aluminium chloride ($AlCl_3$) is a dimer (Al_2Cl_6) in the vapour.

Diphosphane (Disphosphine P_2H_4) A yellow liquid that can be condensed out from phosphine in a freezing mixture. It ignites spontaneously in air.

Disintegration in formation of soil This is a mechanical process in the formation of soils.

Discordant drainage A pattern of drainage that bears no relation to the structure and dip of the underlying rock. The drainage pattern developed on rock strain that have since been removed by erosion–a process known as superimposition, and secondly the drainage pattern was already present before the period of uplift folding that formed the present structure. As the uplift took place the rivers were able to cut down at approximately the same rate and so maintain their courses. The process is known as antecedence.

Dismembered drainage A pattern of drainage in which streams that were once tributaries of a single river system now form separate streams flowing into the sea individually. This results from a rise in seal level, or submergence of the land, which has drawn the lower part of the river system.

Disodium tetraborate 10 water (borax, Na_2B_4)–IOH$_2$O) A white crystalline solid, sparingly soluble in cold water but readily soluble in hot water. It occurs naturally as salt deposits in dry lake beds, especially in California.

Disperse phase *See* colloid.

Displacement It is the shortest distance between two positions of the moving body. When a body moves from A to B, the direction of displacement is AB.

Displacement reaction A chemical reaction in which an atom or group displaces another atom or group from a molecule. A common example is the displacement of hydrogen from acids by metals

$$Zn + 2HCl - ZnCl_2 + H_2$$

Disproportionation A chemical reaction in which there is simultaneous oxidation and reduction of the same compound.

Dissection In geomorphology, the process by which a land surface is cut into by rivers. The term is generally used descriptively.

Dissociation Breakdown of a molecule into two molecules, atoms, radicals, or ions. Often the reaction is reversible, as in the ionic dissociation of weak acid in water.

$$CH_3COOH + H_2O = CH_3COO^- + H_3O^+$$

D horizon In soils, the solid unweathered rock below the soil profile. Some D horizons are unrelated to solid development above. For example, a calcareous bedrock over which acidic drift deposits generate podzols.

Distillation The process of boiling a liquid and condensing the vapour. Distillation is used to purify liquids or to separate components of a liquid mixture. *See* also destructive distillation, fractional distillation, steam distillation, vacuum distillation.

Doab A low-lying area of land composed of alluvium that lies between two converging rivers, chiefly in the Indo-Gangetic plain of N India, *e.g.* Sind Saga Doab lying between the Indus and Jehlam rivers. The term is also used for such areas elsewhere.

Dolomite A mineral consisting of calcium, magnesium carbonates $CaMg(CO_3)_2$. It is also a sedimentary rock that resembles limestone.

Double bond A covalent bond between two atoms that includes two pairs of electrons, one pair being the single bond equivalent (the sigma pair) and the other forming an additional bond, and the other forming an additional bond, the pi bond. It is conventionally represented by two lines, for example $H_2C = O$. *See* multiple bond, orbital.

Double salt When equivalent quantities of certain salts are mixed in aqueous solution and the solution evaporated, a salt may form *e.g.* $FeSO_4 (NH_4)_2 . SO_4 6H_2O$. In aqueous solution the salt behaves as a mixture of the two individuals. These salts are called double salts to distinguish them from complex salts, which yield complex ions in solution.

Drainage basin An area of the Earth's surface drained by a single river system and bounded by a watershed, which separates it from adjoining drainage basins.

Drainage pattern The arrangement of streams within a drainage basin. This reflects the relief and the geological structure of the area, *e.g.* the arrangement of rocks, folds, and faults, and is also influenced by past and present climatic conditions.

Drilling The method by which valuable materials can be obtained from beneath the earth's surface by boring a hole. A number of important minerals and chemicals are obtained by this method *e.g.* oil, natural gas, salt, and sulphur.

Dulong and Petit's law The molar thermal capacity of a solid element is approximately equal to 3R, where R is the gas constant

($25JK^{-1}mol^{-1}$). The law applies only to elements with simple crystal structures at normal temperatures. At lower temperatures, the molar heat capacity falls with decreasing temperature (it is proportional to 3). Molar thermal capacity was formerly called atomic heat–the product of the atomic weight and specific thermal capacity.

Dumas' method A method for determining the vapour densities of volatile liquids. The method utilizes a Dumas bulb; *i.e.* a glass bulb with a narrow entrance tube. The bulb is weighed 'empty' (*i.e.* full of air) then the sample is introduced and the bulb immersed in a heating bath so that the sample boils and expels all the air. When the surplus vapour has been expelled, the bulb is sealed off, cooled, dried, and weighed.

Dumpy level An instrument used in surveying to determine height. It consists of a spirit level and a telescope that are fixed so they can only move horizontally. *See* levelling.

Dune A mount or ridge of sand found in desert and some coastal areas, resulting from the deposition of particles of sand that have been transported by the wind. The major control over the growth of a sand dune is the availability of a continuous supply of sand. The majority of dunes in deserts are either crescentic (branchans) or elongated and narrow (seifs), their precise from being dependent on the direction and strength of the dominant winds.

Duralumin (Trademark) A strong light-weight aluminium alloy containing 3–4 per cent copper, with small amounts of magnesium, manganese, and sometimes silicon. It is widely used in aircraft bodies.

Dust bowl A semiarid area in which the removal of vegetation (*e.g.* by drought, deforestation, or overgrazing) has resulted in extensive wind erosion of the topsoil.

Dyke A wall-like intrusion of igneous rock formed as magma rose up through a near-vertical crack, forcing the rock apart. As a magma cooled it formed a vertical sheet of rock with parallel sides cutting across bedding planes. A dyke can vary in thickness from a few centimetres to hundred of metres, but 2–6m is the average. Dykes tend to occur in large numbers, known as swarms, *e.g.*

on the coasty of the Isle of Arran and on the Isle of Mull, Scotland, from where some of the dykes can be traced across S. Scotland into N. England. *See* also sill.

E

Earth The third planet of the solar system and the only one on which life is known to exist. It is a flattened sphere in shape with an equatorial diameter of 12756 km, and a polar diameter of 12714 km. It makes one rotation on its axis every 23 hours 56 minutes, and orbits the sun in 365 days, 6 hours, and 8 minutes.

Earthflow A mass movement of water-logged material down a slope, while usually leaves bowl-shaped scar on the slope from which it has become detached. The material is deposited as a tongue at the foot of the slope. *See* also mass movement.

Earth movement The movement of a section of the crust of the Earth both horizontally and vertically, including the relative movement of land and sea, movement along faults, and the folding of rocks.

Earth pillar A tall column of soft materials, capped by a boulder, which can reach heights of about 10 m. It is formed by the erosion of the surrounding material by rain water or stream action, while the soil immediately beneath the boulder remains protected. Earth pillars are most frequently found in mountainous areas.

Earthquake A series of shocks that result from sudden earth movement, *e.g.* along a fault, which can be recognized by the passage of earthquake (seismic) waves. The waves spread out from the seismic focus of the earthquake and can cause widespread destruction. In area of the world that lie along the edge of the Pacific Ocean, earthquakes are very common. The severity of earthquakes is measured on the Richter scale.

Earth science Any of the specialized science that study the Earth. These include geography, geology, geophysics, seismology, hydrology, oceanography, biogeography, and pedology (soil science).

Earthworms Two species are common–(1) *Lumbrius terristris,* and (2) *Allolobophora caligionosa.* They are important as soil creatures because (*a*) They digest the organic matter of the soil and leave it back in the soil in the finer stage (*b*) The holes left in the soil serve to increase aeration and drainage, important considerations of soil from the lower depths to the surface, resulting in a mass mixing effect. Of the large organisms, the influence of this creature is perhaps in general the most important. Besides translocating and mixing the soil, these animals pass large quantities through their bodies. The earth is thus subjected to certain digestive processes, with considerable influence on its physical and chemical nature. Though they are about the same size the former is darker and redder, the latter being pale and somewhat pinkish and there is also a difference in mouth parts. Also the girdles of the organisms are located somewhat differently in respects to the segments.

Ecology *See* plant science.

Ecosphere The parts of earth in which life forms can exist, *i.e.*, the land surface, soil, lakes, ocean and the lower atmosphere.

Ecosystem A community of plants and animals sharing a given environment together with nonliving (abiotic) habital that they occupy. Ecosystems form the base units for ecological study. The concept can be applied on a wide range of scale, *e.g.* to the whole world (global ecosystem), tundra, tropical rainforests or a small unit such as a pond or a lake. Ecosystems characteristically derive their energy from sunlight and receive inputs of nutrients, water and, gases, heat, oxygen, carbon dioxide and organic compounds form the outputs. The organisms typically comprise producers (green plants), which convert inorganic compounds through photosynthesis; consumers which feed on plants and each other and decomposers (*e.g.* bacteria, fungi) which promote decay. Ecosystems are thus self-sustaining and mutually independent.

Edaphology Edaphos refers to the soil or earth as a foothold for higher plants. Since all animal and human food comes from directly or indirectly for the soil, crop production knowledge is very much in demand. Hence soil study, in the final analysis, must yield information of an agricultural and economic nature. Edaphology is an examination of soil from the standpoint of higher plants, covers and scientific aspect of soil study.

Effect of drainage on soil temperature The low temperature in soil is due partially to evaporation, and partially to high specific heat. Marked differences have been noticed in the temperatures of drained and undrained soils at any time.

Effect of lime on soil Liming affects the soils in three ways–(1) physically, (2) chemically, and (3) biologically.

Effect of lime on flocculation Hydrogen and calcium cations are more effective perhaps because, of the related electrical potential, of the particles, their degree of hydration, and their migration velocities.

Effect of local factors in soil formation These factors reflect in a great measure. Effects of some local factors, such as the parent material or the slope of the land (topography) and age of the soil. *See* intrazonal soils.

Effect of radiation on soils Direct radiation from the sun, the heat generated by the decomposition of organic matter in the soil, and the heat from the earth's interior are the chief sources of soil heat. Direct heat from the sun plays the major role. The decomposition of organic matter is a slow process.

Effect of tillage on granulation Addition of lime is effective as a granulation agent largely through its influence on biotic forces.

Effervescent soils Many parts of the arid and semi-arid give rise to saline and alkaline effervescences and are known under various names *viz.* reh, kallar and usar. The efflorescences are mainly salts of Na, Ca, and Mg. Soils impregnated with these salts are rendered infertile.

Efficiency It is the ratio between the useful work done by the machine and the total work done one the machine.

Efficiency of soil solutes Transfer of solutes to the top layers of the soil as a result of capillary transference of solutions from lower layers, in places under irrigation by canal water and those in which the sub-soil water table is high as in coastal tracts, Bombay-Deccan. Madras and those near the sea coast. Saline soils of "Solonchak" contain free sodium and other *salts* whereas alkali soils or "Solonchak" do not contain any free salts but sufficient exchangeable Na are found in the irrigated tracts as Bombay-Deccan. The degraded alkali or "solodi" soil is rarely found in India due perhaps to the presence of Ca salts in the saline soil.

Efficiency of bacteria in soils In the absence of the nitrogen fixing bacteria, no nodules will be formed by the leguminous plants and no atmosphere nitrogen will be fixed: It has been found that nodule organisms belong to several distinct strains, each of which is capable of producing nodules on a specific group of legumes.

Efficiency symbol n A measure used for processes of energy transfer; the ratio of the useful energy produced by as system or device to the energy input.

Efflorescence The process in which a crystalline hydrates solid loses water or crystallization to the air. A powdery deposit is gradually formed.

Electrochemical equivalent symbol z The mass of an element released from a solution of its ion when a current of one ampere flows for one second during electrolysis.

Electrochemical series (electromotive series) A series giving the activities of metals for reactions that involve ions in solution. In decreasing order of activity, the series is

K, Na, Ca, Mg, Al, Zn, Fe

Pb, H, Cu, Mg, Ag, Pt, Au

Any member of the series will displace ions of a lower member from solution. For example, zinc metal will displace Cu^{2+} ions:

$$Zn(s) + Cu^{2+}(aq) \rightarrow Zn^{2+}(aq) + Cu(s)$$

Zinc has a greater tendency than copper to form positive ions in solution. Similarly, metals above hydrogen displace hydrogen form acids:

$$Zn + 2HCl \rightarrow ZnCl_2 + Hz$$

The series is based on electrode potentials, which measure the tendency to form positive ions. The series is one of increasing electrode potential for half cell of the type M^{n+} 1 M. Thus, copper (E^{θ} for Cu^{2+} 1 Cu = +0.34V) is lower than zinc (E^{θ} for Zn^{2+} 1 Zn = –0.76 V). The hydrogen half cell has a value $E^{\theta} = 0$.

Electrochemistry The study of the formation and behaviour of ions in solutions. It includes electrolysis and the generation of electricity by chemical reactions in cells.

Electrode Any part of an electrical device of system that emits or collects electrons or other charge carriers. An electrode may also be used to deflect charged particles by the action of the electrostatic field that it produces.

Electrode position The process of depositing a layer of solid (metal) on an electrode by electrolysis. Positive ions in solution gain electrons at the cathode and are deposited as atoms. Copper, for instance, can be deposited on a metal cathode from copper sulphate solution.

Electrode potential A measure of the tendency of an element to form ions in solution.

Electrolytic refining A method of purifying metals by electrolysis. Copper is purified by making the impure metal the anode in an electrolytic cell with copper sulphate electrolytes.

Electron An elementary particle of negative charge (–1.602 192 C) and rest mass 9.109 558 kg. Electrons are present in all atoms in shells around the nucleus.

Electron affinity symbol A Energy released when an atom (or molecule or group) gains an electron in the gas phase to form a negative ion. It is thus the energy of:

$$A + e^{-} - A$$

A positive value of A (often in electron-volts) indicates that heat is given out. Often the molar enthalpy is given for this process of electron attachment (H). Here the units are joules per mole

(Jmol^{-1}), and, by the usual convention, a negative value indicates that energy is released.

Electron-deficient compounds Compounds in which the number of electrons available for bonding is insufficient for the bonds to consist of conventional two-electron covalent bonds. Diborane, B_2H_6, is an example in which each boron atom has two terminal hydrogen atoms bound by conventional electron-pair bonds and in addition the molecule has two hydrogen atoms bridging are only two electrons for the bonding orbital.

Electron diffraction A technique used to determine the structure of substances, principally the shapes of molecules in the gaseous phase. A beam of electrons directed through a gas at low pressure produces a series of concentric rings on a photographic plate. The dimensions of these rings are related to the interatomic distances in the molecules. *See* also X-ray diffraction.

Electrophoresis The migration of electrically charged particles towards oppositely charged electrodes in solution under an electric field–the positive particles to the cathode and negative particles to the anode. The rate of migration varies with molecular size and shape. The technique can be used to separate or analyse mixtures (*e.g.* of proteins). Wetted filter or starch gel is used in the techniques. It can be carried out in conjunction with paper chromatography.

Electroplating The process of coating a solid surface with a layer of metal by means of electrolysis (*i.e.* by electrode-position).

Electrovalent bond (ionic bond) A binding force between the ions in compounds in which the ions are formed by complete transfer of electrons from one element to another element or radical. For example, Na + Cl becomes $Na^+ + Cl^-$. The electrovalent bond arises from the excess of the net attractive force between the ions of opposite charge over the net repulsive force between ions of like charge.

Element A substance that cannot be decomposed into more simple substances. The atoms of an element all have the same proton number (and thus the same number of electrons, which determines the chemical activity).

Elevation The height of a point above a given level (usually mean sea level).

Elevation of boiling point A colligative property of solutions in which the boiling point of a solution is raised relative to that of the pure solvent. The elevation is directly proportional to the number of solution molecules introduced rather than to any specific aspect of the solute composition. The proportionality constant k_B is called the boiling-point elevation constant or sometimes the ebulioscopic constant.

Eluvial In study of soil profile, the A group or eluvial lies at the surface and includes all of the horizons from which material is being removed by leaching.

Elution The removal of an absorbed substance in chromatography column or ion-exchange column using a solvent (eluent). The chromatography column can selectively absorb one or more components from the mixture.

Eluviation The down-washing of clay and other fine particles in free-draining soils. The upper horizons are thus depleted of fine clay particles, becoming an eluvial zone. The particles may be re-deposited at a lower horizon or may be washed out of the soil altogether. Compare illuviation. *See* also leaching.

Emergence The exposure of previously submerged land brought about by a fail in sea level or by the uplift of a continental land mass.

Empirical formula The formula of a compound showing the simplest ratio of the atoms present. The empirical formula is the formula obtained by experimental analysis of a compound and it can only be related to a molecular formula if the molecular weight is known.

Encouragement of granulation in soils In fact the granulation of a clay soil cannot be promoted adequately without the presence of a certain amount of humus the maintenance of organic matter therefore, is of great practical importance, not only chemically and biologically but also physically.

Endogenetic (Endogenic) Denoting the processes and materials that originate with the Earth *e.g.* earthquakes and igneous rocks. Compare exogenic.

Energy acre-furrow slice The amount of organic residue of soil carrier from 150,000,000 to 180,000,000 calories of potential energy, equivalent in heat value to 20 to 25 tons of anthracite coal.

Energy level One of the discrete energies that an atom or molecule, for instance can have according to quantum theory. Thus is an atom there are certain definite orbits that the electrons can be in corresponding to definite electronic energy levels of the atoms. Similarly, a vibrating of rotating molecule can have discrete vibrational and rotational energy levels.

Energy profile A diagram that traces the changes in the energy of a system during the course of a reaction. Energy profiles are obtained by plotting the potential energy of the reacting particles against the reaction coordinate. To obtain the reaction coordinate, the energy of the total interacting system is plotted against position for the molecules. The reaction coordinate is the pathway for which the energy is a minimum.

Energy released by the decay process Organic matter contains considerable potential energy, a large proportion of which is readily transferable to other latent forms or is liberated as hat and outside the soil. Plant tissue, such as heat entering the soil has a heat value of 4 to 5 kilocalories to the gram containing 5000 lbs of dry matter, would mean an addition of 9.000,000 to 11.000,000 kilocalories.

Energy salt *See* magnesium sulphate.

Energy symbol W A property of a system; a measure of its capacity to do work. Energy and work have the same unit the joule(J). It is convenient to divide energy into kinetic energy (energy of motion) and potential energy ('stored' energy). Names are given to many different forms of energy (chemical, electrical, nuclear, etc.); the only real difference lies in the system under discussion. For example, chemical energy is the kinetic and potential energies of electrons in a chemical compound.

Equalization of soil heat Moisture is one of the major factors in respect to the heat capacity of a soil and hence has much to do with its rate both of making up and cooling off.

Equatorial rainforest Also called tropical rainforests. A type of vegetation dominated by very tall trees (60m) that grow in the equatorial regions. The plant life is rich, varied and quick growing in response to the climatic conditions of heavy all round rainfall (above 1500mm constantly) high temperature (25–30°C).

Equipment for soil survey The following is a list of equipment necessary for soil survey. Besides this, it is the general experience of the field workers that unless a quick transport like jeep is provided in the soil survey organisation, the output of work is likely to be small: Post-hole and screw augers to Surveyor's umbrella reach required depth.

Munsell colour chart	Camera. HCl bottle, indictors for
Compass	pH measurement, etc. water bottle.
Abney level	
Tape, khurpi, spade, pick-axe, etc.	For collecting Macromonolith:
Bags, labels, etc., for collecting soil samples.	(i) Chisel, hammer, knife etc. (ii) Monolith farms, 5 ft. long x 4 inches broad x 2 inches deep of wood.
Necessary base maps (topo-sheets, cadastral survey maps, etc.) and field note books.	
Drawing materials and stationary hands lens or magnifying glass for Identifying structure.	For collecting Micromonolith: (i) Standard kit box containing cutter, plunger, blotting paper, Stiff card board, mounting board paste, vinylite
Soil monolith kits.	with spray Arrangements etc.
Plan table with accessories, surveyor's chain, optical square, pegs and survey rods.	(ii) 24 Monolith framers: 10 inches long x 1/2 inch broad x 1 1/2 inches deep, made of plywood.
High-way level with accessories. Tent.	

Equilibrium The condition of a natural system of ecosystem when it is in a state of balance.

Equivalence point The point in a titration at which the reactants have been added in equivalent proportions, so that there is no excess of either. It differs slightly from the end point, which is the observed point of complete reaction, because of the effect of the indicator, errors, etc.

Equivalent weight A measure of 'combining power' formerly used in calculations for chemical reactions. The equivalent weight of an element is the number of grams that could combine with or displace one gram of hydrogen (or 8 grams of oxygen or 35.5 grams of chlorine). It is the atomic weight divided by the valency.

For a compound the equivalent weight depends on the reaction considered. An acid, for instance, in acid-base reactions has an equivalent weight equal to its molecular weight divided by the number of acidic hydrogen atoms.

Erosion The wearing away of the land surface by agents that invoice the transportation of material, the agents of erosion-water, ice, and wind are only capable of having a marked effect on the land surface, when they are transporting fragments of weathered material. In a strict sense, erosion is distinct from denudation in that erosion does not include weathering.

Eruption The process by which material from within the Earth's crust is forced upwards in the form of solids, liquids, or gases, and extruded onto the surface as a result of volcanic activity. *See* also fissure eruption, volcano.

Essential elements from soil solids Nitrogen, phosphorus, potassium, calcium, magnesium, sulphur are used in large quantities from soil solids while iron, manganese, boron, copper and zinc are used by plants supplied in relatively small quantities through the soil solids.

Essential elements' ionic forms in soils The ionic forms are present as of Nitrogen–NH_4^{++}, NO_2^-, NO_3^-; of Potash K^+; of Phosphorus–$PO_4^=$, HPO_4^-, $H_2PO_4^-$; Calcium as a Ca^{++}; Magnesium as Mg^{++}; Sulphur as SO_4^-, SO_3^-.

Essential elements of the soil While all the nutrient elements supplied by the soil must be present in soils and available so that plants may grow normally, six have received the major attention. They are nitrogen, phosphorus, potassium, calcium, magnesium, and sulphur.

Essential elements' transference to available forms Soil is such a complex body that it is very difficult to picture in any detail the various changes that normally occur. Some of the more important transformations do indicate in a very general way of the various changes, especially they affect the availability of the elements that may be limiting factors in plant growth. The productivity of soil certainly depends on the trend of these reactions.

Essential elements used in trace amounts The nutrient elements as iron, manganese, boron, copper and zinc, are used by higher plants in but small amounts and as a consequence are sometimes called trace-elements. Such a designation does not mean that they are less essential than the so-called primary elements.

Essential soil water Higher plants obtain most of their carbon and oxygen, directly from the air by photosynthesis and respiration respectively. Hydrogen is derived, either directly or indirectly from water of the soil all of the other essential elements, except certain supplies of nitrogen acquired from the soil and air indirectly by legumes are obtained from soil solids.

Ester A compound formed by reaction of a carboxylic acid with an alcohol RCOOH + HOR_1–$RCOOR_1$ + H_2O glycerides are esters of long-chain fatty acids and glycerol. If the acid is a carboxylic acid, the formula is RCOOR. Where R and R_1 are organic groups. Esters of low molecular weight acids and alcohols are volatile fragrant compounds. Esters of some long chain carboxylic acids occur naturally in fats and oils.

Ethanol (acetaldehyde, CH_3CHO) A water-soluble liquid aldehyde used as a starting material in the manufacture of several other compounds.

Ethane (C_2H_6) A gaseous alkane obtained either from the gaseous fraction of crude oil or by the 'cracking' of heavier fractions. Ethane is the second member of the homologous series of alkenes.

Ethene (Ethylene, C_2H_4) A gaseous alkene. Ethene is not normally present in the gaseous fraction of crude oil but can be obtained cracking. This is the principal industrial source. The compound is important as a starting material in the organic-chemicals industry (*e.g.* in the manufacture of ethanol) and as the starting material for the production of polythene. Ethene is the first member of the homologous series of alkenes.

Ether A type of organic compound containing the group-o-simple others have the formula R-O-R, where R and R′ are alkyl or aryl groups, which may or may not be the same. They are either gases or very volatile liquids, are very flammable. The common example is ethoxyethane (diethylether, $C_2H_5OC_2H_5$) used formerly as an anaesthetic. Ethers now find application as solvents.

Evaporation The process whereby a liquid changes in a gas. Evaporation from the surface of the earth together with the transpiration (release of water vapour from plants) provides nearly all the water vapour in the atmosphere, this is very important as it is the water vapour that it forms the basis for most types of weather. The rate of evaporation is dependent on the air, temperature, vapour pressure (the amount of vapour already present in atmosphere), the nature of the surface and the wind. Evaporation is most effective when temperatures are high when there is a wind blowing and where there is a large water surface. The highest evaporation rates occur in desert areas and lowest in regions with equatorial climate and cool climate.

Evaporation control methods (1) Two types of mulches (artificial and natural), (2) the paper mulch, and (3) the proper tillage.

Evaporation due to crops With transpiration of water by weeds and surface evaporation, both of these losses effect water that might otherwise be used readily by the crop, and thus the competition is direct and often keen.

Evaporation effect on temperature If the temperature of soil is raised, the capacity of the air for water vapour would increase and evaporation with its cooling effect would again occur. But if the temperature of the system should be lowered, the capacity of the air for vapour would be decreased and condensation, *i.e.* a faster return of molecules to the liquid than the outgo therefrom, would take place.

Evaporation losses' magnitude from soil Some of the runoff and much of the percolation are considered necessary in that such water is superfluous. Some of this removal occurs at a time when it not required by plants. Actually there is not a big competition between the removal of drainage water and the plant, except insofar as nutrients are involved. Surface runoff, however, removes some water that might well go into the soil, and erosion always means a loss of valuable nutrients.

Evaporation of water from soil surface This takes place almost entirely at the soil surface, exceptions being where large cracks occur that allow the thermal loss directly from the subsoil.

Evaporimeter (atmometer) An instrument used to measure the amount and rate of evaporation of a liquid. There are two main types; one measures the loss in weight of a known quantity of liquid and another measures the change in level of the surface of the liquid.

Evaporite A sedimentary rock that has been formed by the precipitation of minerals from a concentrated solution, which have been dried out by evaporation, *e.g.* rock salt (sodium chloride) and gypsum. This occurs chiefly under desert conditions where there is prolonged evaporation from a shallow stretch of water.

Excess of essential elements Too much nitrogen in proportion to the phosphorus available to plants may encourage too much calcium, may interfere with phosphorous and boron nutrition or may encourage chlorosis due to a reduction in the availability of the soil iron, zinc or manganese. An under supply of calcium may retard both nitrogen and phosphorus availability.

Exchangeable calcium and magnesium If these are lost by leaching, the acidity of the soil gradually increase can be said in anyone soil. There is a definite correlation between the pH and the activity of these constituents.

Exfoliation (Onion weathering) A weathering process in which the outer layers of a rock split off into thin concentric sheets or scales. It has been suggested that this results from insulation weathering where solar heating during the day is followed by rapid-night cooling. The expansion and contraction resulting from this wide diurnal range of temperature would cause stresses in the rock leading to the exfoliation.

Exothermic Denoting a chemical reaction in which heat is evolved (*i.e.* heat flows from the system or the temperature rises). Combustion is an example of an exothermic process. Compare endothermic.

F

Factors affecting soil temperature *See* soil temperature.

Factors influencing evaporation of water from soil The principal conditions are (1) relative humidity of atmosphere, (2) the amount of water at the soil surface especially its rate of replacement, (3) the temperature difference of the two moisture phases, and (4) wind velocity.

Factors influencing flow of gravity water (1) the colloidal matter may clog not only the macropores but also the smaller connecting channels, thus increasing the pressure necessary to ensure complete drainage, (2) heavy soils that crack during dry weather at first allow rapid percolation of rain water. Later however, these cracks may swell shut thus reducing percolation to a minimum. The constriction may be clogged somewhat by adsorbed water molecules due to hydration of pores. (3) Entrapped air, which is sure to be abundantly present in such a complicated system, may also seriously obstruct the rate of percolation.

Factors influencing soil character These depend on the climatic conditions of the place and the soil material already present there.

Fahrenheit scale A temperature scale on which 32° represents the freezing point of water and 212° represents the boiling-point of water.

Fans When streams descend from uplands, a sudden change in a gradient sometimes occurs as the stream emerges at the lower level. A deposition of sediment is thereby forced, giving rise to alluvial fans. They differ from deltas in their location and in the character of their debris. Fan material is generally gravely and stony, more or less porous and in general, well drained.

Faraday constant *See* Faraday's laws.

Faraday symbol–F A unit of electric charge equal to the charge required to discharge one more of a singly charged ion. One faraday is $9.648\,670 \times 10^4$ coulombs. *See* Faraday's laws.

Faraday's laws (of electrolysis) Two laws resulting from the work of Mechael Faraday on electrolysis:

1. The amount of chemical change produced is proportional to the electric charge passed.
2. The amount of chemical change produced by a given charge depends on the ion concerned. More strictly it is proportional to the relative ionic mass of the ion, and the charge on it. Thus the charge Q needed to deposit m grams of an ion of relative ionic mass, M, carrying a charge z is given by:

$$Q = Fmz/M$$

i.e. $9.648\,670 \times 10^4$ coulombs.

Fatty acid *See* carboxylic acid.

Fehling's solution A freshly mixed solution used for testing for the presence of reducing sugars and aldehydes in solution. When boiled with equal amounts of Fehling's A [copper (II) sulphate solution] and Fehling's B (sodium potassium tartrate and sodium hydroxide solution) reducing sugar and aldehydes produce a brick red precipitate of copper (I) oxide, and copper metal is produced. The tartrate is present to complex with the original copper (II) ions to prevent precipitation of copper (II) hydroxide.

Feldspar A group of minerals that consist of silicates of aluminium, combined with potassium, sodium, calcium and barium. It forms one of the most common minerals in igneous rocks and is the most abundant mineral in the Earth's crust. It is divided into two groups; alkali feldspars, containing either sodium or calcium or both in varying proportions. The crystals are white, pink or grey. The type of feldspar present is important in the classification of igneous rock.

Fen An area of low-lying marshy land that is underlain by a peaty soil. Fens are characterised by soils that are rich in humus and are usually alkaline. When drained the dark fenland soils are very fertile. Drainage of ten results in shrinkage and without proper management, can lead to a formation of a dry dust that can be removed by high winds.

Ferallitic soils A group of red coloured soils that occur on the old plantation surfaces of the humid tropics formed under forests. Heavy rainfall causes leaching to great depths. The red colour derives from iron compounds. The loamy texture and fertility can easily be impaired by agricultural use.

Fermentation A chemical reaction produced by micro-organisms (moulds, bacteria or yeasts). A common example is the formation of ethanol from sugars.

$$C_6H_{12}O_6 - 2C_2H_5OH + 2CO_2$$

Fermentation layer In soils a layer of partly under-composed litter lying below the freshly fallen organic material and above plant remains that have been fully decomposed to form humus. This stratified structure is typical of morhumus.

Ferredoxins A group of red-brown proteins of green plants, blue-green algae, and some bacteria. They contain non-haem iron in association with sulphur at the active site. They are strong reducing agents (very negative redox potentials) and function as electron carries, for example in photosynthesis and nitrogen fixation. Ferric chloride, *See* iron (III) chloride.

Ferrel's law A law that states as a result of the Earth's rotation, a body moving in any direction over the Earth's surface will end to be deflected, to the right in the N hemisphere and to the left in the S hemisphere. The effects of this are to be seen most markedly in the movements of air and water. *See* also coriolis force (Named after the US meteorologist William Ferrel who first postulated the law in 1859).

Ferric oxide *See* iron (III) oxide.

Ferric sulphate *See* iron (III) sulphate.

Ferrisoils A group of red-coloured soils occurring in the humid tropics that resemble ferrallitic soils in structure and development, but are less intensely leached and weathered. They are formed where surface erosion is constantly removing weathered material from the surface (*e.g.* on slopes) they are more fertile than ferrallitic soils.

Fertallitic soils A group of red-coloured soils that occur on the old plantation surfaces of the humid-tropics formed under forests. Heavy rainfall causes strong leaching to great depths. The red colour derives from iron compounds. The loamy texture and fertility can be easily impaired by agricultural use.

Field capacity *See* water holding capacity of soil.

Filtration The process removing suspended particles from a fluid by passing or forcing the fluid through a porous material (the filter). In laboratory filtration, filter paper or dintered glass is commonly used.

Final products of soil decomposition processes.

Complex	*Simple*
Humus–a colloidal	Carbon dioxide
Complex in which the	Carbon dioxide
lignoproteinate is	Nitrate
considered important	Sulphates
	Phosphates
	Ca compounds

Fine earths The mineral particles of the soil may be classified according to size under four heads:

Very coarse Visible to naked eye, stone and gravel.

Coarse Visible to the naked eye, sands of various sizes.

Fine Visible under ordinary microscope silt particles.

Very Fine Visible satisfactorily only. Clay particles under electron microscope

Highly colloidal.

The last three grades are the most desirable as components of arable soils predominate. They are spoken as fine earths.

Fine organic chemicals Carbon compounds, such as pesticides, dyes and drugs, produced only in small quantities. Their main requirement is that they must have a high degree of purity, often as high as 95 per cent. They are manufactured for special purposes, *e.g.* for use in spectroscopy, pharmacology and electronics.

Fire clay A soft fossil clay that often occurs below coal seams. It may have been the soil in which the swamp plants that eventually formed the coal seam grew. Fire clay can withstand great heat (upto 1600°C) and is used in the manufacture of refractory bricks for the linings of furnaces.

Firedamp methane occurring in coal mines Explosions can occur if quantities of methane accumulates. The air left after an explosion is called black-damp. After-damp is the poisonous carbon monoxide formed.

Fissile Denoting the tendency of some rocks and minerals such as mica, slate and shale, to split (cleave) along defined planes. *See* also cleavage.

Fixation and release of phosphorus in soil This is of special importance in soils. Phosphorus is fixed in the soil by adsorption, double decomposition and isomorphic replacement. Phosphate fixation is positively correlated with ion and aluminium iron in soil, which react with soluble phosphates to form both crystalline and amorphous precipitates.

Flame test A preliminary test in qualitative analysis in which a small sample of a chemical is introduced into a non-luminous bunsen flame on a clean platinum wire. The flame vaporizes part of the sample and excites some of the atoms, which emit light at wavelengths characteristic of the metallic elements in the sample.

Flint A hard dark-grey form of silica that occurs in chalk as scattered nodule or irregular shape, tabular sheets or vertical pipes. Although it is very hard and brittle and thus enabled to work by Stone Age man into tool. It is used as building material in areas where it occurs.

Flocculating capacity The coagulating capacity of various cations is in the order of H and Ca, Mg, K, Na. The colloidal complexes of humid region soils as are dominated by calcium and hydrogen, and therefore such soils gradually tend to assume a coagulated condition, in the filed which is of some consequences in respect to granulation.

Flocculation It is coagulation of mater in a colloidal state. It is a special character of most clay soils. If a colloidal clay suspension is treated by a little calcium hydroxide, the tiny clay particles almost immediately coalesce into floccules, and that because of their combined weight, sink to the bottom of the container, leaving the supernatant liquid clear. This phenomenon is flocculation because of the peculiar appearance of the aggregates.

Flocculation effect on granulation When flocculation occurs, the colloidal matter, especially that of a highly viscous and gelatinous nature, is coagulated and a localization occurs. This disrupts the soil mass and tends to encourage the crumb structure, which usually does not provide for the stabilization aggregates.

Floodplain The low-lying land that borders a river and is subjected to periodic flooding. It is composed of deposits of sediments (alluvium) of variable thickness laid down by the flood waters above the rock floor and is bounded by low bluffs.

Floods and drainage channels The individual farmer can lessen the damage due to floods to some extent by keeping the drainage channels of his fields open and in good repair. However, aforestation of upper catchment areas of rivers, construction river embankments and execution of multipurpose river valley projects as preventive measures call for governmental action.

Flood tide The rising tide, which raises the water level from the low-water mark to the high water mark along coastlines and in estuaries. The flood tide does not necessarily cause flooding, although exceptionally high tides may do so.

Flourescein A fluorescent dye used as an absorption indicator.

Fluorescence The absorption of energy by atoms, molecules, etc. followed by immediate emission of electromagnetic radiation as the particles make transitions to lower energy states.

Fluoridation The introduction of small quantities of fluoride compounds into the water supply as a public-health measure to reduce the incidence of tooth decay in children.

Fluoride *See* halide.

Fluorine A slightly greenish-yellow highly reactive gaseous element belonging to the halogens (group VII of the periodic table). It occurs notable as fluorspar (CaF_2) and cryloite (Na_3AlF_3) but traces are also widely distributed with other minerals. It is slightly more abundant than chlorine, accounting for about 0.065 per cent of the Earth's crust.

Symbol F; b.p. –118xC; m.p. –223xC; d. 1.696 kgm^{3}; p.n. 9; r.a.m. 18.99840.

Fluorocarbon A compound derived from a hydrocarbon by replacing hydrogen atoms with fluorine atoms. Fluorocarbons are unreactive and most are stable upto high temperatures. They have a variety of users-in aerosol propellants, oils and greases, and synthetic polymers such as PTFE.

Fluospar A naturally occurring form of calcium fluoride, FAF_2.

Folic acid (pteroylglutmic acid) One of the water-soluble B-group of vitamins. The principal dietary sources of folic acid are leafy vegetables, liver and kidney. Deficiency of the vitamin exhibits itself in anaemia in a similar manner to vitamin B deficiency. Folic acid is important in metabolism in various coenzyme forms, all of which are specifically concerned with the transfer and utilization of the single carbon (C_1) group. Before functioning in this manner, folic acid must be reduced to either dihydrofolic acid (FH_2) or tetrahydrofolic acid (FH_4). It is important in the growth and reproduction of cells, participating in the synthesis of pureneoo and thymine. *See also* vitamin B complex.

Foot-pound-second system In this system of units, foot, pound and second are the units of length, mass and time respectively.

Force It is a push or pull, which when acting on a body, changes or tends to changes its state of rest or of uniform motion in a straight line.

Force exerted by freezing The force developed by freezing of water is equivalent to 150 tons to the square foot, an almost irresistible force.

Formaldehyde The spray is done with fromaline while disinfecting soil and the seeds. Useful to control the damping off disease.

Formation of soils Soils are formed by the mechanical disintegration chemical decomposition of parent rocks. The breaking up of rocks (*i.e.*, weathering) may thus be physical or chemical weathering. The principal agents of physical weathering are wind, water and gravity.

Formic acid (methanoic acid) The most simple carboxylic acid, HCOOH. It is present in ant bites and contributes to the nettle sting reaction.

Form or grade of lime That should be used depending on (*i*) its relative cost, (*ii*) fineness of grinding, (*iii*) rapidity of action, and (*iv*) convenience of handling. The calcium and magnesium content of these substances is usually reported in terms of calcium and magnesium oxide and not of the elements themselves.

Forms of calcium in mineral soils Original minerals such feldspar, hornblende and calcite; as primary nutrient elements.

Forms of essential elements absorbed by plants The nutrient element s get into the form of a dynamic biological medium, the soil solution. Many of the salts projected into it is the processes are those of high ionization. As a consequence, the soil solutions, especially that of a humid region, is in fact ionic, the degree depending in part on the concentration. When the solution is dilute, considerable proportion of the soluble constituents are present as ions. As the solution becomes more concentrated during a dry spell, the proportion of the soluble substances in molecular association tends to increase. This holds good for arid-regions also.

Forms of lime From the chemical point, it refers to one and only one compound-calcium oxide. Thus used it includes all compounds of calcium and magnesium employed kin a practical way, nor only to raise the pH of an acid soil but especially to rectify its physiological conditions.

Forms of nutrient elements in soil The soil nutrients presented to the plants by soil are both in ionic and molecular forms, the proportion depending on conditions. In an arable soil to predominate, not only because of the nature of soil solution, but also because of the greater ease with which the ions penetrate the absorbing tissue.

Forms of primary nutrient elements occurring in soils An examination of a soil in a general way the nutrient elements exist in two conditions (1) complex and insoluble forms, (2) simple compounds usually soluble in the soil water and readily available to the higher plants.

Formula A representation of a chemical compound using symbols for the atoms and subscript numbers to show the numbers of atoms present. *See* empirical formula, molecular formula, structural formula.

Forward reading vernier If n V.D. are equal to (n 1) main scale divisions, it is called forward reading Vernier.

Fossil 1. The remains of, or impressions left in rocks by, long dead animals and plants. Most fossils consists of hard skeletal material because soft tissues and organs rot away very quickly. Mineral salts from surrounding rocks gradually replace the hard termed petrifaction. Alternatively the organic material dissolves away leaving and impression or mauled in the surrounding rocks. Trace fossils (*e.g.* of dinosaur footprints) provide indirect evidence of prehistoric life forms.

2. The remains up a plant or animal preserved in sedimentary rock. These may be the whole or part of the organism itself, such as a bone, a piece of shell, a seed or a leaf, which has usually been chemically altered and replaced by mineral matter or a trace or impression of the organism, *e.g.* a wormcast or a footprint of an animal. Rocks of a similar age and depositional environment contain similar fossils and therefore the fossil content can be used to date and correlate layers of rock. The term can also be applied to a feature that has been buried by geological processes, such as an erosion surface.

Fraction A mixture of liquids with similar boiling points collected by fractional distillation.

Fractional crystallization Crystallization of one component from a mixture in solution. When two or more substances are present in a liquid in solution, on cooling it a certain temperature, one substance will preferentially form crystals, leaving the other substance in the liquid (or dissolved) state. Fractional crystallization can thus be used to purify or separate substances if the correct conditions are known.

Fractional distillation (fractionation) A distillation carried out with partial reflex, using a long vertical column.

Free radical An atom or group of atoms with a single unpaired electron. Free radicals are produced by breaking a covalent bond; for example:

$$CH_3Cl \rightarrow CH_3 + Cl$$

They are often formed in light-induced reactions. Free radical are extremely reactive and can only be stabilized and isolated under special conditions.

Freezing This has a great practical importance in alerting the physical condition of field soils, especially if they are heavy in texture.

Freezing of water The force developed by freezing of water is equivalent to about 150 tons to the square foot, an almost irresistible process.

Friable 1. Denoting a rock or mineral that is easily crumbled and disintegrated when pressed between the fingers, *e.g.* kaolin.

2. Denoting a soil with a well-defined crumb structure.

Friction When a body is moving or sliding over another surface, its motion is opposed by a force, called the force of friction.

Fructose ($C_6H_{12}O_6$) A sugar found in fruit juices, honey and cane sugar. It is a ketohexose, existing a pyranose form when free. In combination (*e.g.* in sucrose), it exists in the furanose form.

Fuel cell A type of cell in which fuel is converted directly into electricity. In one form, hydrogen gas and oxygen gas are fed to the surfaces of two porous nickel electrodes immersed in potassium hydroxide solution. The oxygen reacts to form hydroxyl (OH^-) ions, which it releases into the solution, leaving a positive charge on the electrode.

Fuller's earth A natural clay used as an absorbent and industrial catalyst.

Fumaric acid An unsaturated dicarboxylic acid, which occurs in many plants. The fumarate ion participates in several important metabolic pathways, *e.g.* the Krebs cycle, purine pathways and the urea cycle.

Fundamental units The units which are independent and cannot be converted into each other are called Fundamental Units. The units of length, mass and time are three such units.

Furnose A sugar that has a five-membered ring (four carbon atoms and one oxygen atom). *See* also sugar.

Fused Describing a solid that has been malted and solidified into a single mass. Fused silica, for example, is produced by melting sand.

Fused ring *See* ring.

Fusiform initial *See* initial.

G

Galvanic cell *See* cell.

Galactose ($C_6H_{12}O_6$) A sugar found in lactose and many polysaccharides. It is an aldohexose, isomeric with glucose. *See* also sugar.

Galvanizing A process for coating steel with zinc by dipping in a bath of molten zinc, or by electro-deposition. The zinc protects the steel from corrosion.

Gamma rays Streams of gamma radiation.

Gas The state of matter in which forces of attraction between the particles of a substance are small. The particles have freedom of movement and gases, therefore, have not fixed shape or volume.

Gas laws Laws relating to the temperature, pressure and volume of a fixed mass of gas. The main gas laws are Boyle's law and Charles' law.

Boyle's and Charles' laws can be combined into an equation of state for ideal gases:

$$pV_m = RT$$

where, V_m is the molar volume and R the molar gas constant. For n moles of gas:

$$PV = nRT$$

All real gases deviate to some extent from the gas laws, which are applicable only to idealized systems of particles of negligible volume with no intermolecular forces. There are several modified equations of state that give a better description of the behaviour of real gases, the best known being the van der Waals equation.

Gaseous component of soil The gaseous component of the soil occupies a portion of the pore space. Water and gases can replace each other. If a soil becomes waterlogged, the plant roots and micro-organisms are deprived of oxygen, which is so essential for their growth and well-being.

Gas for ripening of fruits To quicken the ripening process in fruits, they are taken from trees after they have attained full maturity and placed in gas chambers to ripen rapidly.

Gas natural The gaseous form of hydrocarbons, consisting chiefly of methane, that is found in underground reservoirs in the Earth's crust, often above deposits of petroleum. Natural gas is formed under the same conditions as oil and originate from the decomposition of organic matter. Also *See* crop science.

Gasoline *See* petroleum.

Genesis of organic deposits Marshes bogs and swamps provide the conditions suitable for the accumulation of organic deposits.

Genesis of silicate clays These clays develop most readily from silicate minerals like feldspar, mica and the others. The process begins with hydrolysis which is a sense, base exchange. The initial step is:

$$KAlSi_3O_8 + HOH = HAlSi_3O_8 + KOH$$

Genesis of soils In the first place the disruption of the original silicate minerals pave the way for clay formation. Gradually this highly colloidal material is synthesised. The time finally comes when living, organisms decisively intervene, organic matter accumulates, especially at the surface and the whole trend of weathering is sharply modified.

Geology The study of the structure and composition of the earth. It is divided into a number of branches; cosmology–the study of evolution of the earth in space: geophysics–the study of earth processes: geochemistry–the study of the distribution of minerals and how the rocks are formed: mineralogy and petrology–the study of rocks, mineral deposits ores: palaeontology–the study of fossils: stratigraphy–the study of succession of rocks and fossils in geological changes over in geological time in surface relief and climate.

Geological time The division of time since the formation of each and the relationship of this division to the formation of rocks. Geological time is divided into a hierarchy of time intervals in descending order, these are eras, periods, epochs, ages.

Geometric mean The geometric man of n numbers is the nth root of their product. Thus the geometric mean of the five numbers 1, 2, 3, 4, 5 is $\sqrt{(1.2.3.4.5)}$–2.605.

Geomorphology The study and interpretation of the origins and development to landforms on the earth's surface. It includes the study of the landforms themselves and of the processes creating the landforms. The chief areas covered are weathering, hillslopes and mass movement, fluvial features, aeolian features, glacial and periglacial features, coastlines, and karst landforms.

Geophysics The study of the physics of the Earth, both the crust and the interior, and of the mechanisms and causes of movement (*e.g.* faulting and folding) in the crust also includes the application of physical methods for example the surveying techniques used to search for petroleum, natural gas and other minerals of economic importance.

German silver *See* nickel-silver.

Glacial These are formed by the grinding action of ice and brought down by glaciers in the upper regions of Himalayas; found in Dehradun region and parts of Kashmir.

Glacial aeolian deposits Such deposits other than loss do occur such as abode, volcanic ash and sand dunes. Soils from volcanic ash are light and porous. Sand dunes are always changing place and thus are not important agriculturally.

Glacial deposits During the pleistocene, northern and central Europe and parts of Asia invested and invaded by a succession of great ice sheets. There was a interglacial period which is proved by the presence of forest beds and accumulation of organic matter between the glacial deposits of different ages and by evidence of prolonged erosion of the successive layers of glacial drift. Some of the interglacial intervals were times of warm or even semitropical climate in regions that are now definitely temperate.

Glacial lake deposits This group presents a wide variation in physical and chemical characteristics as the soil materials of any of the physiographic area. Being deposited by weather, they have been subject to much sorting and stratification and range from coarse gravels to fine clays. There is also much variation in respect to lime content, pH, colour and structure.

Glacial soils *See* moraines.

Glaciation The covering of an area with an ice sheet or by a glacier due to a lowering of temperatures or the processes and landforms resulting from these conditions.

It is also a period in the past during which extensive areas were covered by ice *e.g.* quaternary glaciation. This is also called ice age.

Glass A hard transparent material made by heating calcium oxide (lime), sodium carbonate and sand (silicon (IV) OXIDE). This produces a calcium silicate–the normal type of glass, called soda glass.

Gley soil (glei soil) An intrazonal hydromorphic soil that develops in waterlogged conditions. This may be, for example, because an impervious layer prevents infiltration of rainwater, or in low-lying land subject to flooding. Gleying may affect the whole soil profile or result in agley horizon–a compact layer of bluish-grey structureless clay.

Glucose (dextrose, grape sugar) A monosaccharide ($C_6H_{12}O_6$) occurring widely in nature as D-glucose. It occurs as glucose units in glycogen (hydrolysis to glucose) and sucrose (hydrolysis to glucose and fructose) as well as in starch and cellulose. Its significance to mammals lies in its participation in energy-storage and energy-release systems. Glucose is made available by hydrolysis of starch in foods initially by salivary amylase and later by pancreatic amylase to give maltose (2-glucose units), then by intestinal maltase to glucose itself. *See* also sugar.

Glyceride An ester formed between glycerol (propane-1,2,3-triol) and one or more carboxylic acids. Glycerol has three alcohol groups and if all three groups have formed esters, the compound is a triglyceride. Naturally occurring fats and oils are triglycerides of long-chain carboxylic acids (hence the name 'fatty acid'). The main carboxylic acids occurring in fats and oils are: octadecanoic acid (stearic acid), a saturated acid $CH_3(CH_3(CH_2)_{16}$ COOH. cis-9-octadenenoic acid (oleic acid) and unsaturated acid. $CH_3(CH_2)$-CH (CH_2)-COOH.

Glycerine *See* propane-1,2,3-triol.

Glycerol (Glycerine; 1,2,3 propanetriol) It is an alcohol with three HO groups. Glycerol is biologically important as the alcohol involved in lipid formation (these particular lipids being called glycerides).

These glycerides are the fat stores of the body and enter into metabolic pathways, particularly when carbohydrate levels are low. Digestion of lipids occurs in the small intestine after being emulsified under the influence of bile salts. They are then converted back to the fatty acids by pancreatic lipase, (*i.e.* the reverse of the reaction above) and enter metabolic pathways via reaction with coenzyme A.

Gycine *See* amino acids.

Gneiss A coarse-grained metamorphic rock with a banded structure. It is formed by the large-scale application of heat and pressure associated with mountain building and volcanic activity.

Gold Gold is used in jewellery often alloyed with copper. Pure gold is 24 carat; 9 carrat indicates that 9 parts in 24 consist of gold. Symbol: Au; m.p. 1063°C; b.p. 2710°C; r.d. 19.3; p.n. 79; r.a.m. 196.97. Gold is a rare native metal that is yellow in colour, does not dissolve in acid, and is soft enough to be cut with a knife. It is found in two main types of deposit; (1) in placer deposits, particularly in river beds, derived from the weathering of gold-bearing rock, and (2) in hydrothermal veins, where it is associated with quartz and hyrite (fool's gold).

Gradient 1. The steepness of a sloping surface, which may be expressed either by the angle in degrees from the horizontal or by representing the vertical distance and the horizontal distance as a ratio as follows: average gradient between two points = horizontal distance/difference in height.

2. *See* pressure gradient.

Graham's law (of diffusion) The principle that gases diffuse at a rate that is inversely proportional to the square root of their density. Light molecules diffuse faster than heavy molecules. The principle is used in the separation of isotopes.

Grajite It is composed of various compounds of SiO_2, $(Al_2)O_3$, Fe_2O_3 CaO, MgO, K_2O, P_2O_5.

Gram equivalent The equivalent weight of a substance in grams (gramme). Symbol: g a unit of mass in the metric system, defined as 10^{-3} kilogram.

Gram Symbol g A unit of mass defined as 10^{-3} kilogram.

Gram-molecule *See* mole.

Granite A light coarse-grained igneous rock consisting of quartz, feldspar (mainly alkali), and usually mica, with a variety of other minerals in small quantities. It is acid in composition, which results from its high silica content (over 70 per cent). Granites are intrusive rocks and occur in many forms such as sills, dykes, and batholiths. They are hard and resistant and occur as large masses in uplands, which have been exposed by denudation; for example, the tors of Dartmoor are weathered out of the granite batholith that forms the moor.

Granular All round aggregates may be placed in this category. The term more properly refers to those not over ½ inch in diameter. When the granules are specially porous, the term crumb is used. A granular structure is characteristic of many surface soils, especially those high in organic matter and is the physical condition of particular importance in land under cultivation.

Granular disintegration A weathering process that involves the breaking down of rock into individual grains. Both mechanical and chemical weathering processes may be responsible for disintegration.

Granulation A process for enlarging particles to improve the flow properties of solid reactants and products in chemical processes.

The larger a particle, and the freer from fine materials in a solid, the more easily will it flow.

Granule genesis Several interrelated factors account for granule formation. Of these (1) wetting and drying (2) freezing and thawing (3) physical activities of roots and small animals (4) influence of decaying organic matter and of the slimes from the micro-organisms and other forms of life that accomplish decomposition and (5) the modifying effects of adsorbed bases.

Grape sugar *See* dextrose $C_6H_{12}O_6$.

Graphite An allotrope of carbon. Graphite is a good conductor of heat and electricity. The atoms are arranged in layers. Graphite is used as a solid lubricant.

Also called plumbago, black lead, a soft black crystalline form of pure carbon, found in many metamorphic rocks, especially metamorphosed coals and other carbonaceous sediments. It occurs in veins and as lens shaped masses. Graphite is used in paint, rubber, and lead pencils, as a lubricant, for making metallurgical crucibles, in electroplating, as electrodes for electrical furnaces, and as a moderator in nuclear reactors.

Gravel A deposit of unconsolidated material, ranging in size from 2 to 60 mm (some authorities give 2–10 mm). In size it lies immediate above sand and below cobbles. The term is, however, frequently used for any loose material in the size range 2–20 mm. The particles are usually water worn and hence rounded, and are derived from more than one type of rock.

Gravimetric analysis A method of quantitative analysis in which the final analytic measurement is made by weighing.

Gravimetric method for soil water measurement Soil moisture is usually determined by taking a sample of soil in the field by means of any auger or tube, and drying it in an oven at 105°C. The loss of water on drying may be represented in percentage on dry weight of soil or in acre-inches on volume basis. The latter mode of expression helps to understand the relation of soil to water as well as air.

Gravitational constant is equal to the force of attraction between two bodies of unit masses placed at a unit distance from each other.

Gravitational units of work (1) gm-O-n = 1 gm wt × 1 cm = 981 ergs.

(2) Kg-metre = 1 kg wt × 1 metre = 9.81 joules.

(3) Ft lb = 1 lb wt × 1 ft = 32 poundals.

Gravitational water Having important energy relationship of soil and plants to this water even though it is superfluous or excess in nature, it receives too much attention for two reasons (1) the necessity of adequate drainage and (2) the loss of plants nutrients that its removal entails.

Gravity A force of attraction between the earth and a body.

Gravity flow of water Soil has two kinds of pore-micropores and macropores. The former are usually small enough to create a viscous friction of such magnitude as to greatly retard the downward movement of gravitational water unless subjected to considerable pressure. But if the interstices are not too small, the weight of water accumulating above or below may be sufficient to force passage. The macropores on the other hand, are larger enough to facilitate a very ready movement of water in response to gravity unless impeded in some way.

Great soil group of world Two major groups are named as (1) Zonal humid and (2) Zonal aridic sols. From these various other sub-groups emerge.

Grey-brown podzolic soil A zonal soil of cool moist climates that is intermediate in character between true podzols and brown earths. The soil is less leached and hence is slightly less acid than podzols and as a result contain more soil fauna, which incorporate humus into the A horizon. The B horizon generally has a brown black structure and an accumulation of alluvial clay.

Grey desert soil *See* sierozem.

Grey podzolic soils A truly podzolic soil is designated as the region of grey or typical podzol soils because of the characteristic colour of the A_2 horizon. It may be said also that these soil materials are so drastically leached during soil formation and the upper layers of soil left. A coarse-grained sandstone in which the particles are either angular or of unequal size. The term is also used as a proper name for Millstone Grit.

Group 1. In the periodic table, a series of chemically similar elements that have similar electronic configurations. A group is thus a column of the periodic table. For example, the alkali metals, all of which have outer S^1 configurations, are called group 1.

2. (Functional group) In organic chemistry, an arrangement of atoms that bestows a particular type of property on a molecule and enables it to be placed in a particular class, *e.g.* the aldhyde group –CHO. *See* also functional group.

Group O elements *See* noble gases.

Group I elements *See* alkali metals.

Group II elements *See* alkaline-earth metals.

Group III elements A group of elements in the periodic table, often divided into the main group (sub-group IIIA) boron (B), aluminium (Al), gallium (Ga), indium (In) and thallium (TI), and the B sub-group, scandium (SC) yttrium (Y), lanthanum (La), and actinium (AC). The B sub-group division is to some extent not helpful as scandium, yttrium and lanthanum may be conveniently considered in the transition metals etc.

Group IV elements A group of elements in the periodic table; divided into the main group (sub-group IVA) consisting of carbon (C), silicon (Si), germanium (Ge), tin (Sn), and lead (Pb); and sub-group IVB; titanium (Ti), zirconium (Zr), and hafnium (Hf), which are transition metals.

Group V elements A group of elements of the periodic table sometimes called the chalcogens. They are oxygen (O), sulphur (S), selenium (Se), tellurium (Te), and polonium (Po). The electronic configurations are all s^2p^4 with no vacant inner orbitals. These configurations are all just two electrons short of an inert gas structure, consequently they have high electron affinities and are almost entirely non-metallic in character.

Gully An incised water-worn channel, which is particularly common in semiarid areas. It is formed when overland flow down a slope, especially following heavy rainfall, is concentrated into rills, which merge and enlarge into the gully. Gully erosion is the erosion of soil and rock by the concentration of runoff into gullies. Large volumes of soil and debris may be removed, especially where the soils and underlying rock are soft. Removal of protective vegetation and ploughing down-slope can also contribute to this form of erosion. Its effects can be prevented by contour ploughing. Gully erosion occurred on a large scale in the Mid-west of the USA during the 1920s when large areas were ploughed up and the vegetation removed for agriculture.

Gums These complex carbohydrates are found in different kind of plants and are of various kinds. They are formed as a result of the breakdown of cell wall and cell contents. This phenomenon is known as gumosis. They are insoluble in alcohols. Some of the common gums are of gum arabica from *Acacia senegal*, gum urens from *Sterclaria urens*, gugul from *Bosswellia serrata* and gum guar from *Cymopsis tetragonoloba*.

Gun cotton *See* cellulose nitrate.

Gunmetal *See* bronze.

Gunpowder A powdered mixture of sulphur, charcoal, potassium and nitrate used as an explosive. *See* calcium sulphate.

H

Haber process An important industrial process for the manufacture of ammonia, which is used for fertilizers and for making nitric acid. The reaction is the equilibrium:

$$N_2 + 3H_2 = RNH_3$$

The nitrogen used is obtained by fractionation of liquid air and the hydrogen by the oxidation of hydrocarbons (from natural gas).

Haematite (ferric iron oxide, Fe_2O_3) The principal ore of iron, containing over 70 per cent iron. It occurs chiefly in sedimentary rocks, in which it is derived from altered iron carbonates and silicates and produced the red coloration in sandstones. Hematite occurs in a number of different forms such as crystalline (specular iron ore), powdery (red ochre) and in the form of lumps (kidney ore).

Hafnium A transition metal found in zirconium ores. Hafnium is difficult to work and can burn in air. It is used in nuclear reactors. Symbol Hf; m.p. 2220°C; b.p. 5230°C, r.d. 13.3; p.n. 72; r.a. m. 178.49.

Hail Precipitation in the form of pellet of ice (hillstones) that develop in and fall from a cumulonimbus cloud, either as a cold front or where intense heating of the surface causes rapidly ascending convection currents. Also *See* Crop Science.

Half-life (half-life period) Symbol T, The time taken for half the nuclei of a sample of a radioactive nuclide to decay. The half-life of a nuclide is a measure of its stability (Stable nuclei can be thought of as having infinitely long half-lives). If N_0 is the original number of nuclei, the number remaining at the end of half-life is $N_0/2$, at the end of two half-lives in $N_0/4$ etc.

Halide A compound containing a halogen. The inorganic halides of electropositive elements contain ions of the type Na^+Cl^- (sodium chloride) or K^+Br^- (potassium bromide). Transition metal halides often have some covalent bonding. Non-metal halides are covalent compounds, which are usually volatile. Examples are carbon tetrachloride (CCl_4) and silicon tetrachloride ($SiCl_4$). Halides are named as bromides, chlorides, fluorides or iodides.

Halite *See* rock salt.

Haloalkane (Alkyl halide) A type of organic compound in which one or more hydrogen atoms of an alkane have been replaced by halogen atoms. Haloalkanes can be made by direct reaction of the alkane with a halogen.

Halogenation A reaction in which a halogen atom is introduced into a molecule. Halogenations are specified as chlorinations, brominations, fluorinations, etc.

Halogens A group of elements (group VII of the periodic table) consisting of fluorine, chlorine, bromine, iodine and the short-lived element astatine. The halogens all have outer valence shells that are one electron short of a noble-gas configuration. Because of this, the halogens are characterized by high electron affinities and high electro negativities of the halogens favour the formation of both uninegative ions.

Halosere A series of successional stages leading to a climax originating in a saline area. *See* also sere, succession.

Halloysite *See* kaolin

Hardening (of oils) A process for converting liquid vegetable oil into solid or semi-solid fats. Oils are glycerides of mainly unsaturated carboxylic acids. Hydrogenation of the double carbon-carbon bonds using a nickel catalyst produces saturated glycerides, which have higher melting points. Hydrogenation of vegetable oils is used in the manufacture of margarine.

Hardness (of water) Hard water is water that will not readily form a lather with soap owing to the presence of dissolved calcium, iron and magnesium compounds. Such compounds can react with soap to produce insoluble salts, which collect as solid scum. The effectiveness of the cleansing solution is thus reduced.

Hardness of water is of two types, temporary hardness and permanent hardness. Only temporary hardness can be removed by boiling the water. Hardness of water is usually expressed by assuming that all the hardness is due to dissolved calcium carbonate, which is present as ions. It can be estimated by titration with a standard soap solution or with edta. *See* also soap, water softening.

Hardness scale (Mohs' scale) A scale devised by the mineralogist Friedrich Mohs in 1812 as a guide to the hardness of minerals. The determination of hardness is one of the most important tests in the identification of minerals.

Hardpan A hard compact layer at the surface of the soil or at a lower horizon within the soil profile. It may consist of concentration of particles transported by soil solution or chemical compound precipitation from solution. *See* also claypan, iron plan, duricrust.

Headward erosion The erosion at the source of a stream in an upstream direction. Headward erosion frequently results from spring sapping in which water undercuts the slope from which the spring flows. This leads to the development of subsurface pipes and tunnels, which eventually become exposed as streams as they cave in. As the source of a stream recedes under headward erosion it may breach the watershed and capture streams from other side. *See* also river capture.

Hard water Water that contains calcium, magnesium and iron salts in solution. Hard water forms scales in kettles and hot water pipes and does not easily form a lather when used with soap. Water that runs off chalk and limestone tends to be hard as it contains salts derived from the solution of rocks.

Heat Energy transferred as a result of a temperature difference. The term is often loosely used to mean internal energy of the particles.

It is common in chemistry to define such quantities as heat of combustion, heat of neutralization etc. These are in fact molar enthalpies for the change given the symbol H_M. The unit is usually the kilojoule per mole (KJ mol^{-1}). By convention, H is negative for an exothermic reaction. Molar enthalpy changes stated for chemical reactions are changes for standard conditions, which are 298K (25°C) and 101 325 Pa (1 atmosphere).

Heat balance The state of equilibrium that exists of Earth between incoming radiation, mainly from the Sun and outgoing re-radiation and reflection from the Earth. It is a world-wide average as flow-latitude areas receive more radiation that they lose, while the opposite, occurs nearer the poles, the heat necessary to preserve the balance being transferred by the movement of air masses and ocean currents.

Heat exchangers Devices that enable the heat from a hot fluid to be transferred to a cool fluid without allowing them to come into contact. The normal arrangement is for one of the fluids to flow in a coiled tube through a jacket containing the second fluid. Both the cooling and heating effect may be of benefit in conserving the energy used in a chemical plant and in controlling the process.

Heat in soil formation Differential expansion and contraction of minerals in the rocks in temperature or cold regions set up internal tensions and gradually tears them apart. High average temperatures also expedite chemical weathering more especially under humid condition.

Heat in soil Terrestrial bodies are continually giving off energy waves into the atmosphere and since they are mostly of the infra red type and make no impression on the eye they are often spoken as dark rays. Their energy capacity is high and their emission is contiguous, much heat is on the way. This accounts in part, especially during clear weather for rapid changes in surface-soil temperatures.

Heat lost by radiation For soil, *See* soil temperature fluctuations.

Heat of atomization The energy required in dissociating one mole of a substance into atoms. *See* heat.

Heat of combustion The energy liberated with one mole of a substance burns in excess oxygen. *See* heat.

Heat of formation The energy change when one mole of a substance is formed from its elements.

Heat of neutralization The energy liberated when one mole of an acid or base is neutralized.

Heat of reaction The energy change when molar amounts of given substances react completely.

Heat of solution The energy change when one mole of a substance is dissolved in a given solvent to infinite dilution (in practice, to form a dilute solution).

Heavy hydrogen *See* deuterium. A naturally occurring stable isotope of hydrogen kin which the nucleus contains one proton and one neutron. The atomic mass is thus approximately twice that of H.

Heavy water Deuterium oxide, D_2O.

Height of water rise The height to which water rises in a capillary tube is expressed by the formula:

$$h = \frac{2T}{rdg}$$

where, T is surface tension in dynes per centimetre, r the radius of tube in centimetres, d the density of the liquid and g is gravity in dynes per centimetre. The smaller is the radius of the tube, the higher will be rise of the liquid in it.

Helium It has the electronic configuration $1s^2$ and consists of a nucleus of two protons and two neutrons (equivalent to a particle) with two extra-nuclear electrons. It has an extremely high ionization potential and is completely resistant to chemical attack of any sort. The gas accounts for only 5.2×10^{-4} per cent of the atmosphere; upto 7 per cent occurs in some natural gas deposits. Helium is the second most abundant element in the universe. Helium also has an isotope. It is formed in nuclear reactions and by decay of tritium. A colourless monatomic gas; the first member of the noble gases (group O of the periodic table).

Hematite It is a soil mineral Fe_2O_3. It is clay mineral of whose disintegration takes place thus:

$$2Fe_2O_3 - 3H_2O_3 2Fe_2 O_3.3H_2O$$

Limonite yellow

$$4Fe - 3H_2O + O_2 = 2Fe_2O_3 - 3H_2O$$

Ferrous oxide Limonite

Hestidine Hornblend is composed of silicate of calcium, magnesium and iron with sodium silicate.

Heterotrophic Most soil bacteria are heterotrophic *i.e.* their energy and carbon come directly from the soil organic matter. The general purpose decay and ammonifying bacteria belong to this group.

Heterotrophism is a type of nutrition in which the principal source of carbon is organic, *e.g.* the organism cannot make all its own organic requirements from inorganic materials. Most heterotrophic organisms are chemotrophic (*i.e.* show chemoheterotrophism); these comprise all animals and fungi and most photographic (*i.e.* show photohetrotrophism). The non sulphur purple bacteria, for instance, require organic molecules such as ethanol and acetate. Compare autotrophism. *See* chemotrophism, phototropism.

Hexagonal crystal *See* crystal system.

Hexane (C_6H_{14}) A liquid alkane obtained from the light fraction of crude oil. The principal use of hexane is as a solvent.

Hexose A sugar that has six carbon atoms in its molecule. *See* sugar.

Hexose monophosphate shunt *See* pentose phosphate pathway.

Hexyle group The group of $CsH_{11}CH_2$ having a straight chain of carbon atoms.

Higher plants and soil reaction A very acid soil will present low pH, low exchangeable calcium and magnesium, soluble aluminium, iron, maganese, possible toxic organic matter and a low availability, if not an actual lack of nitrogen and phosphorus. An intermediate physiological set up is presented by a moderately to slightly acid soil, which seems for the average plant to be quite satisfactory since the chemical and biological agencies are well in advance.

H-ions heterogeneity of soil solutions A solution is not the one like an ordinary homogenous solution. The solute concentration of a soil solution, due to adsorption, is heterogeneous. Therefore, it is correct to visualize the H-ions being concentrated near the colloidal interfaces, are becoming less numerous as the outer portions of the films are approached.

Homocrylic compound A compound containing a ring made up of the same atoms. Benzene is an example of homocrylic compound. Compare heterocyclic compound.

Homogenous relating to a single phase A homogeneous mixture, for instance, consists of only one phase.

Homologous series A group of organic compounds possessing the same functional group and having a regular structural pattern so that each member of the series differs from the next one by a fixed number of atoms.

Humic acid An organic acid that is formed when water passes slowly through a mass of decaying vegetation (humus). The acid is effective in the chemical decomposition of rocks and formation of soils. The efficiency is aided to the action of the humus in relating the acid so that minerals are subject to decomposition over long periods. Mosses and lichens on rock surfaces may develop small saucer-shaped depressions under them that are the result of weathering by humic acid.

Humic micelles These like the lamellar particles of clay, exhibit the double layer phenomenon and carry a swarm of adsorbed cations in exactly way. The same exchangeable cations (Ca, H^-, Mg^{++}, K^+, Na^+ etc.) are found in the outer layers and base exchangeable properties of a higher order occur.

Humid region soils These are broadly grouped under three heads (1) cold or frigid (2) temperature and (3) tropical and sub tropical. And according to climate these are Tundra, Podzolic and Lateritic. They are also identified according to the vegetation growing on them which is different for each of them.

Humus It may be defined as a complex and resistant mixture of brown and dark brown amorphous and colloidal substances that develop synthetically as the original organic tissues suffer enzymic dissolution by the various soil organisms. It is a natural body and though exceedingly variable and heterogeneous, it possesses properties that distinguish it sharply from other natural organic aggregates.

Humus as controlled by soil nitrogen As the nitrogen, especially of humid region soils, is carried almost wholly by the organic matter, its further consideration is exploration of its relation with the humus of the soil. Both these show variations at various places and under various conditions. One of the economy phase has to do a lot with the control of nitrogen as it exerts lot of influence on humus accumulation of the soil.

Humus' organic matter The quantity of organic matter present varies with the type of soil, the cultivation, tillage, manorial and rotation practices adopted for its management, climatic conditions including temperature and water supply, and the biological pressure. Climate governs the pace and time of the decomposition of organic matter in soil, while natural vegetation determines the quantity and quality of organic material added each year.

Humus of soil It is a highly colloidal but unlike the mineral colloids of the soil, it is amorphous and not crystalline. Its surface exposure, its adsorption, its catalysis are comparatively far in excess of those exhibited by mineral clays. The presence of 1 per cent of humus in mineral soil means an increase in exchange capacity of about 2 milliequivalents, while clay, 1 per cent, can adsorb to the extent of 0.3 to 0.6 of a milliequivalent. A fully synthesised humus of mineral can adsorb moisture from a saturated atmosphere about 8090 per cent while a mineral clay can absorb about 15–20 per cent. The low plasticity and cohesion of humus is a significant practical feature. Since its maintenance is heavy, field soils cannot help but alleviate unfavourable structural characteristics induced by large amounts of clay.

Hydration A weathering process in which water is taken up by the minerals of a rock. This may cause considerable expansion of the minerals. The stresses caused in the rock by alternate wetting and drying may lead to its disintegration. Hydration is particularly effective in clays where it causes the break-up (slaking) of the clay.

Hydration of silicate clay A common case of hydration is calcium sulphate or gypsum ($CaSO_4.2H_2O$). Not only do molecules of many salts carry combined water but cations seem to have the same capacity. Na and K ions are heavily hydrated. Ca and Mg ions carry water molecules but not in the amounts that characterise the alkali ions.

Hydraulic force The erosive force of moving water, which depends for its efficiency upon the column of water involved, the velocity of the water and the nature of the surface on which the force is exerted. The load carried by the water is not involved in producing this force. Hydraulic force is most effective on highly jointed and creviced rock where water impounds and compresses an air pocket. The pressure exerted in this way may gradually break-up the rock. Hydraulic force is an important component of both fluvial and marine erosion.

Hydrazine (N_2H_4) A colourless liquid that can be prepared by the oxidation of ammonia with sodium chlorate (I) or by the gas phase reaction ammonia with chlorine. Hydrazine is a weak base, forming N_2H_4HCl with strong acids. It is a powerful reducing agent, reducing salts of the noble metals to the metal. Anhydrous hydrazine ignites spontaneously in oxygen and reacts violently with oxidizing agents. The aqueous solution, hydrazine hydrate have been used as a fuel for jet engine and for rockets.

Hydrazoic acid (Hydrogen azide, azomide, HN_3) A colourless liquid with a nauseating smell. It is highly poisonous and explodes in the presence of oxygen and oxidizing agents. It can be made by distilling a mixture of sodium azide (NaN_3) and a dilute acid. It is usually used as an aqueous solution. The salts of hydrazoic acid (azides); especially lead azide [$Pb(N_3)_2$], are used in detonators because of their ability to explode when given a mechanical shock.

Hydrazone A type of organic compound containing the C : NNH_2 group, formed by the reaction between an aldehyde or ketone and hydrazine (N_2H_4). Derivatives of hydrazine are often used to produce crystalline products, which have sharp melting points that can be used to characterize the original aldehyde or ketone. Phenylhydrazine ($C_6H_5NH\text{-}NH_2$), for instance, produces phenylhydrazones.

Hydride A compound of hydrogen. Ionic hydrides are formed with highly electro-positive elements and contain the H ion (hydride ion). Non-metals from covalent hydrides, as in methane (CH_4)

or silane (SiH_4). The boron hydrides are electron-deficient covalent compounds. Many transition metals absorb hydrogen to form interstitial hydrides.

Hydrobromic acid (HBr) A colourless liquid produced by adding hydrogen bromide to water. It shows the typical properties of a strong acid and it is a strong reducing agent. A convenient way or producing hydrobromic is to bubble hydrogen sulphide through bromine water. Although it is not as strong as hydrochloric acid, yet dissociates extensively in water and is a good proton donor.

Hydrocarbon A compound of hydrogen and carbon. Different types of hydrocarbons form the economically important deposits of peat, coal, petroleum and natural gas.

Hydrochloric acid (HCl) A colourless fuming liquid made by adding hydrogen chloride to water: $HCl(g) + H_2O(l) - H_3O^+(aq) + Cl^-(aq)$

Dissociation into ions is extensive and hydrochloric acid shows the typical properties of a strong acid. It reacts with carbonate to give carbon dioxide and yields hydrogen when reacted with all but the most unreactive metals. Hydrochloric acid is used in the manufacture of dyes, drugs and photographic materials. It is also used to pickle metals, *i.e.* clean the metal prior to electroplating. Hydrochloric acid denotes protons with ease and is the strongest of the hydroblaic acids. The concentrated acid is oxidized to chlorine by such agents as potassium permanganate and manganese (IV) oxide.

Hydrocortisone (cortisol) A steroid hormone produced by the adrenal cortex having glucocorticoid activity. *See* glucocorticoid.

Hydroelectric power (HEP) Electric power generated from the energy of falling water. The water drives turbines, which in turn drive electricity generators, conditions suitable for a power station site include high annual precipitation; a river supplied with precipitation from a wide catchment area; an existing lake, or a valley narrow enough to dam in order to create a reservoir to guarantee regular water supply during dry periods.

Hydrofluoric acid (HF) A colourless liquid produced by dissolving hydrogen fluoride in water. It is a weak acid, but will dissolve most silicates and hence can be used to etch glass. As the interatomic distance in HF is relatively small, the H-F bond is very high, therefore unable to donate protons easily hydrofluoric acid is a weak acid. *See* bond energy.

Hydrogen An essential element in living tissues. It enters plants, with oxygen, as water and is used in building up complex reduced compounds such as carbohydrates and fats. Water itself is an important medium, making up 70–80 per cent of the weight of organisms, in which chemical reactions of the cell can take place. Hydrogenated compounds, particularly fats, are rich in every and on breakdown release energy for driving living processes.

A colourless gaseous element. Hydrogen has some similarities to both the alkali metals (group I) and the halogens (group VII), but is not normally classified in any particular group of the periodic table. It is the most abundant element in the earth's crust and atmosphere (by pass). It occurs principally in the form of water and petroleum products; traces of molecular hydrogen are found in some natural gases and in the upper atmosphere. The gas may be prepared in the laboratory by the reaction of dilute hydrochloric acid with a metal that lies above hydrogen in the electrochemical series, magnesium and zinc being commonly used.

$$Zn + 2HCl \rightarrow H_2 + ZnCl_2$$

Hydrogen of coal The manufacture of oil from coal with the use of hydrogen. In normal circumstances, oil drilled from the ground is cheaper and more plentiful than oil produced from coal. The process was used by Germany during World War II because of a shortage of crude oil.

Hydrogen azide *See* hydrazoic acid.

Hydrogen bromide (HBr) A colourless sharp-smelling gas that is very soluble in water. It is produced by direct combination of hydrogen and bromine in the presence of a platinum catalyst or by the reaction of phosphorous tribromide with water. It dissolves in water to give hydrobromic acid. Hydrogen bromide is rather. inactive chemically. It will not conduct electricity in the molten state, indicating that it is a molecular compound.

Hydrogen carbonate (bicarbonate) A salt containing the ion $^{-}HCO_3$.

Hydrogen chloride (HCl) A colourless gas that gives a strong irritating odour and fumes strongly in moist air. It is prepared by the action of concentrated sulphuric acid on sodium chloride. The gas is made industrially by burning a stream of hydrogen in chlorine. It is not particularly reactive but will form dense white clouds of ammonium chloride when mixed with ammonia. It is very soluble in water and ionizes almost completely to give hydrochloric acid. It will not ionize and the resultant solution shows no acidic properties. Hydrogen chloride is used in the manufacture of organic chlorine compounds, such as polyvinyl chloride (PVC).

Unlike the other hydrogen halides hydrogen chloride will not dissociate on heating, indicating a strong H-Cl bond. *See* hydrochloric acid.

Hydrogen fluoride (HF) A colourless liquid by the reaction concentrated sulphuric acid on calcium fluoride; CaF_2 (s) + H_2SO_4 (aq) – $CaSO_4$ + 2HF(l). It produces toxic corrosive fumes and dissolves readily in water to give hydrofluoric acid. Hydrogen fluoride is a typical of the hydrogen halides as the individual H-F units are associated into much larger units, forming zigzag chains and rings. This is caused by hydrogen bonds that form between the hydrogen and the highly electronegative fluoride ions. Hydrogen fluoride is used extremely as a catalyst in the petroleum industry. *See* hydrofluoric acid.

Hydrogen peroxide (H_2O_2) A colourless syrupy liquid, usually used in solution in water. On contract with manganese (IV) oxide acting as a catalyst:

$$2H_2O_2 - 2H_2O + O_2$$

Hydrogen peroxide can act as an oxidizing agent with potassium magnate (VII). It is used as a bleach and in rocket fuel. The strength of solution is usually given as volume strength–the volume of oxygen (dm^3) at STP given by decomposition of 1 dm^3 of the solution.

Hydrogen sulphite (sulphuretted hydrogen, H_2S) A colourless very poisonous gas with an odour of bad eggs. Hydrogen sulphide is prepared by reacting hydrochloric acid iron (II) sulphide. It is tested for by mixing with lead nitrate, with which it gives a black precipitate. Its aqueous solution is weakly acidic. Hydrogen sulphide reduces iron (I) (III) chloride to iron (II) chloride, forming hydrochloric acid and a yellow precipitate of sulphur. Hydrogen sulphide precipitates insoluble sulphides and is used in qualitative analysis. It burns with a blue flame in oxygen to form sulphur (IV) oxide and water. North sea gas contains some hydrogen sulphide, which is removed before supply to the consumer.

Hydrolase An enzyme that catalyses a hydrolysis reaction. Digestive enzymes are an example. Hydrolases play an important part in rendering insoluble food material into a soluble form, which can then be transported in solution.

Hydrology The study of water on the earth including its chemical and physical properties, occurrence, distribution and circulation on the surface and below ground.

Hydrolysis A reaction between a compound and water. Some examples are:

Salts of weak acid

$$Na_2CO_2 + 2H_2O - 2NaOH + H_2CO_3$$

Esters

$$CH_3COOC_2H_5 + H_2O = CH_3COOH + C_2H_5OH$$

Certain inorganic halides

$$SiCl_4 + 4H_2O - Si(OH)_4 + 4HCl$$

When salts with strong cations and weak acid radicals, such Na_2CO_3, K_2CO_3, $Ca(HCO_2)_2$, go into solution they undergo hydrolysis and develop alkalinity. Far Na_2CO_3 the reaction is:

$$Na_2CO_3 + 2HOH\ 2NaOH + H_2CO_3$$

Since the dissociation of the NaOH is greater than that of the weak H_2CO_3, a domination of OH ions results. In general, a reaction between compound and water, particularly one involving H^+ or OH^- ions.

Hydrolysis in soils This is essentially an exchange of constituent parts between water with strong bases produces hydroxides of iron, magnesium, calcium, aluminium, etc. Hydrolysis takes place in several successive stages according to drainage and leaching conditions consequent upon topography, rainfall and nature of rock formation.

Hydrolysis in weathering Of the chemical changes, hydrolysis occurs first, especially in the weathering of feldspar, mica and other similar minerals.

Hydrosphere The total water surrounding the earth. This includes not only the surface water (in solid or liquid form) of oceans, seas, lakes, rivers, ice caps etc. but also atmospheric water and water below the earth's surface *e.g.* within the soil and in underground streams.

Hydroxide clays The base exchange of hydroxide clay is exceptionally low, hardly equal to that of the kaolin clays. Two types of clays are found–the silicate clays characteristic of temperature regions and the iron and aluminium hydroxide clays found in the tropics and semitropics. Moreover, the silicate clays are more important because they are more widely spread over the agriculture regions of the world. These soils in general are yellow or red, as there has been adequate opportunity for the hydration and oxidation of some of the iron. Moreover, they are much lower in plasticity and cohesion than the silicate clays.

Hydroxide of lime This form of lime is commonly and referred to as the hydrated. Since it is produced by passing water to burnt lime the hydroxides that results are often spoken of as slaked lime. It is a white powder, highly caustic and not at all pleasant to handle.

Hydro-oxypropanoic acid (lactic acid) *k* A colourless liquid carboxylic acid:

$$CH_3CH\,(OH)\,COOH$$

See optical activity.

Hygrometer An instrument that measures the humidity of the atmosphere. The most common type uses a human hair, which expands or contacts as the relative humidity increases or decreases, while another type uses a lithium chloride wire, the resistance of which varies with the changing humidity. A continuous record, a hydrogram, showing changes of relative humidity of the atmosphere over a given period of time is obtained from a hydrograph, which is a hygrometer fitted with an inked pen resting on a rotating drum. *See* also psychrometer.

Hydroscopic coefficient When any dry porous substance, such as soil sample, is exposed to water vapour, it will take up a certain amount of moisture depending on the nature and magnitude of the surface presented, the temperature and the degree of humidity. The amount of water thus absorbed on the particle surfaces from an atmosphere of known relative humidity and at a standard temperature is designated as the hygroscopic coefficient. It's relating to Wilting Coefficient is:

$$\text{Wilting Coefficient} = \frac{\text{Hygroscopic coefficient}}{0.68}$$

Hygroscopic water This water is tenaciously held by soil. It exists as a very thin film at the solid-liquid interfaces. This layers of water m, molecules, has a maximum thickness that probably does not exceed 4–5 millimicrons. Because of the tremendous pressures exerted upon them, approaching 10,000 atoms. At the interface, the molecules are closely packed and most of the layer is non-liquid. In any case, in the neighbourhood of the hygroscopic coefficient the pressure is so much less, approximately 31 atoms or pF 4.5, that a transition to the liquid state is gradually taking place.

I

Igneous rock Rock that has solidified form molten magma. Igneous rock may be either extrusive or intrusive. Extrusive of the Earth as magma before cooling and are generally glassy or fine-crystalled. Intrusive rocks cool and solidify within the crust of the Earth; these may be either plutonic, which cool deep within the crust and usually have large crystals, or hypabyssal, which cool at intermediate depths and contain moderate-sized crystals. Igneous rocks may also be classified according to their chemical composition. For example, one classification is based on silica content and has the following divisions: acid rocks (over 66 per cent silica), intermediate rocks (55–60 per cent), and ultrabasic rocks (less than 45.5 per cent silica). *See* also metamorphic rock and sedimentary rock.

Illite and montmor illionite Their general formula is $Al_2O_3\ 4SiO_2\ 5H_2O$, and have two sheets of silica to one of alumina. Illite lattices are fixed, but the plate space of montmorillionite can expand after taking up water. They have high base exchange capacity. Montmorillionite type of clay mineral is the most useful in plant growth.

Indian lateritic soils contain largely kaolinitic type clays, alluvial soils have mostly illitic clay minerals and black cotton soils are rich in montomorillionite clays. Thus, a knowledge of the mineralogical character of clay is as necessary as that of clay content for judging the precise behaviour of soil in the field.

Illuviation The deposition in a soil of fine material such as clays, bases and colloidal organic matter that has been washed downwards from above. Such accumulations often form in the B horizon. Compare eluviation.

Illuvial horizon of soil A zone of accumulation (illuviation) especially of iron and aluminium compounds. It is prominent in true podzols as a reddish-brown horizon.

Impermeable Denoting a rock that is non-porous and does not allow water or air to pass through it. Granite is impermeable, although the presence of cracks and fissures in the rocks may make it possible.

Impervious Denoting a rock through which water cannot easily soak. It can apply to porous but fine-grained sediments such as clays and to non-porous rocks such as massive unjointed granite. Compare impermeable.

Indicator A substance used to test for acidity or alkalinity of a solution by a colour change. Examples are litmus and phenolphthalein. Universal indicator shows a range of colour changes over a wide range from acid to alkaline and can be used to estimate the pH.

Indigenous *See* crop science.

Indium A soft silvery metallic element belonging to group III of the periodic table. It is found in minute quantities primarily in zinc ores. It is used in alloys, in several electronic devices and in electroplating.

Inductive effect The effect in which atoms or groups in a compound can attract (–1) or push away electrons (+1), forming polar bonds.

Indurated honey combmass The separation of iron oxide in the form of nodules and the gradual cementation of the latter to form an indurated honey combed mass is characteristic of lateritic soils.

Inert gases *See* noble gases.

Infiltration The seepage of water into the soil. Soil characteristics such as the structure, texture, presence of a vegetation layer, and existing moisture content and the nature, especially intensity of the precipitation influence the rate of percolation. The maximum rate at which rainfall can be absorbed by a soil in a given condition is known as the infiltration capacity (usually expressed as mm per hour). Once this is exceeded the soil quickly becomes saturated and overland flow takes place.

Influence of moisture upon soil temperature May be it is a question of acquisition of insulation; loss of energy to the atmosphere, or the movement of heat to and from within the soil, the percentage of water present is always important. Water regulation is the key to what little temperature control it is possible to exert on field soils.

Infrared (IR) Electromagnetic radiation with longer wavelengths than visible radiation. The wavelength range is approximately 0.7 mm to 1 mm. Many materials transparent to visible light are opaque to infrared, including glass. Rock salt, quartz, germanium, or polyethylene pumps and lenses are suitable for use with infrared. Infrared radiation is produced by movement of charges on the molecular scale; *i.e.* by vibrational or rotational motion of molecules. Of particular importance in chemistry is the absorption spectrum of compounds in the infrared region.

Inorganic chemistry The branch of chemistry concerned with elements other than carbon and with the preparation, properties and reactions of their compounds.

Inorganic components of soils From what is stated generally it would be evident that compounds of silicon, calcium, magnesium, iron, potassium, sodium and aluminium from the principal chemical constituents of soils. Besides these, the soil contains small quantities of a large number of other mineral elements like boron, manganese, molybdenum, zinc, copper, cobalt, iodine and fluorine which are known to be essential for the well being of plants in very small quantities. Such elements are called minor or trace elements.

Input-output analysis The measurement of the relationship between the level of production of a particular set of goods and services (output) and the raw material, labour and other manufacturing inputs required to produce at that level of output. By identifying input-output coefficients, it is possible to trace through the effects of a set of output requirements on the production pattern of the economy. This system of analysis was invented by soviet economist R.H. Kontorovich and developed by W. Leontiel in the 1950s.

Insoluble Describing a compound that has a very low solubility in a specified solvent.

Intake The point at which air or a liquid is channelled into a plant or a mine. In such cases, surface makes allow fresh air to be drawn through an underground working. Some factories have intakes where water from a lake or river is piped into the works cooling or cleaning purposes.

Intensity of gravity If M = mass or earth; R = radius of earth; m = mass of a body; then the force of attraction between the earth and the body is GMm/R^2. This force per unit mass is called intensity of gravity. This can also be represented as:

$$\text{Intensity of Gravity} = \frac{GMm}{R^2} m = \frac{GM}{R^2}$$

This is also called acceleration due to gravity

Intercalation compound A compound that has a structure based on layers and in which there are layers of a different character interleaved in the basic structural units. For example, the micas phlogopite (KmgS $(OH)_2Si_3AlO_{10}$) and muscovite ($KAl_2(OH)_2$ Si_3AlO_{10} are formed by interleaving K ions replacing a quarter of the silicon layers in talc and pyrophyllite respectively. *See* also lamellar compound.

Interflauve The ridge of the higher ground that lies between two rivers in the same drainage system.

Intermediate rock A igneous rock containing 55–66 per cent silica and less than 10 per cent free quartz. It contains some feldspars, which can be either sodic plagioclase or alkali feldspar or both. Examples of intermediate rocks include andesite, diorite and syentie. *See* also acid rock, basic rock, ultrabasic rock.

Intermittent saturation zone A layer of soil that lies just below the surface layer of soil and which will contain ground water after rainfall, but which will tend to dry out during periods of drought. In this way the zone contains water only intermittently and the frequency of rainfall will determine the period during which the zone is saturated.

Intermolecular forces Forces of attraction between molecules rather than forces within the molecule (chemical bonding). If these intermolecular forces are weak the material will be gaseous and as their strength progressively increases materials become progressively liquids and solids. The intermolecular forces are divided into H-bonding forces and van der Waals forces, and the major component is the electrostatic interaction of dipoles.

Internal energy (intrinsic energy) Symbol–U The quantity of energy possessed by a given substance or system. The internal energy is the sum of the kinetic and potential energies of the atoms or molecules of the system. In practice, changes in internal energy are important in chemical reactions. *See* enthalpy.

Intrazonal soil A mature soil that differs from the zonal soil developed under the prevailing climatic conditions as a result of local factors such as relief or parent material. Intrazonal soils include hydromorphic soils (associated with area of poor drainage) calcimorphic soils (which are developed in saline conditions). Compare azonal soil, zonal soil.

Intrazonal soil groups Soils over wide areas vary greatly as to rock formation and geologic origin in general, tend alter a considerable lapse of time to become alike in general characteristics if the climatic influences are reasonably uniform and continuous and if erosion is not a serious factor. Such soil groups are spoken of as zonal–*i.e.* they are of such wide expanse as to be thought of as continental.

Intrusion A solidified mass of igneous rock formed from molten magma that has forced its way into and through pre-existing rocks, especially along lines of weakness and then cooled. *See* also batholith, boss, dyke, sill, stock.

Intrusive rock Igneous rock that has solidified below the earth's surface. Granite is the most common intrusive rock. Compare extrusive rock. *See* hypabyssal rock, plutonic rock. *See* also intrusion.

Inversion A change from one optical isomer to the other. *See* Walden inversion.

Invert sugar *See* sucrose.

Iodeothane (ethyl iodide, C_2H_5I) A colourless liquid alkyl halide made by reaction of ethanol with iodine in the presence of red phosphorous.

Iodine A dark-violet volatile solid element belonging to the halogens (group VII of the periodic table). It occurs in seawater and is concentrated by various marine organisms in the form of iodides. Significant deposits also occur in the form of iodides. The element is conveniently prepared by the oxidation of iodides in acid solution (MnO_2). Industrial methods similar use oxidation of iodides by sulphur dioxide followed by oxidation, depending on the source of the raw materials. Iodine and its compounds are used in chemical synthesis, photography, pharmaceuticals and dye-stuffs manufacture.

Iodine monochloride (ICl) A dark red liquid made by passing chlorine over iodine. It has properties similar to those of its constituents halogens. Iodine monochloride is used as a non aqueous solvent and as an iodating agent in organic reactions.

Iodoform *See* tri-iodomethane.

Iodomethane (methyl iodide, CH_3l) A liquid alkyl halide made by reaction of methanol with iodine in the presence of red phosphorous.

Ions According to the theory of electrolytic dissociation, many compounds when in solution break up into electrically changed portions called ions. Ions may be single atoms or group of atoms. KCl dissociates in to K^+ and Cl^- ions while $NaNO_3$ yields ionize almost completely. The degree of dissociation is in general greater for inorganic salts than for the organic ones.

Ion exchange A process that takes place in certain insoluble materials, which contains ions capable of exchanging with ions in the surrounding medium.

Ionic bond *See* electrovalent bond.

Ionic crystal A crystal composed of ions of two or more elements. The positive and negative ions are arranged in definite pattern and are held to by electrostatic attraction. Common examples are sodium chloride and caesium chloride.

Ionic double layer phenomenon The minute and heterogeneously dispersed clay particles are electrically active, ordinarily carrying a considerable electrical potential. This is due to an ionic double layer phenomenon, the inner is anion and the outer layer is of cations.

Ionic exchange and the availability of nutrients Base exchange functions under two more or less distinct conditions as far as plant roots and micro-organisms are concerned. In one case the nutrient dislodged by the H-ions which affects most of the interchange, find their way into the outer and free portion of the soil solution. There they may contact the absorption surface or root hairs and micro-organisms or react again with the soil solids or be removed in drainage water. The large amounts of calcium and potassium lost by leaching indicate that many ions escape biological absorption in this way.

Ionic forms of nutrients Knowing the soluble salts that commonly enter the soil water, it is easy to list nutrient ions that are brought in contact with the absorbing tissues of micro-organisms and higher plants. They are:

Nitrogen	$NH_4^+NO_2^-, NO_3$
Phosphorus	$PO_4^-, HPO_4^-, H_2PO_4^-$
Potassium	K^+
Iron	Fe^{2++}, Fe^{+++}
Manganese	Mn^{++}, Mn^{++++}
Copper	Cu^+, Cu^{++}
Carbon	$CO_3^=, HCO_3^-$
Calcium	Ca^{++}
Magnesium	Mg^{++}
Sulphur	SO_3^-, SO_4
Zinc	Zn^{++}
Boron	$BO_3^=$
Water	H^+_3, OH^-

Ionic product The product of concentrations:

$$K_w = (H^+)(OH^-)$$

in water as a result of a small amount of self-ionization.

Ionization The process of producing ions. There are several ways in which ions may be formed from atoms or molecules. In certain chemical reactions, ionization occur by transfer of electrons.

Ionization energy *See* ionization potential.

Ionization potential (IP) Symbol–I The energy required to remove an electron from an atom (or molecular or group) in the gas phase, *i.e.* the energy required for the process:

$$M \rightarrow M^{+} + e^{-}$$

It gives a measure of the ability of metals to form positive ions. The second ionization potential is the energy required to remove the electrons and form a doubly charged ion:

$$M \rightarrow M^{2+} + e^{-}$$

Ionization potentials stated in this way are positive; often they are given in electron volts. Ionization energy is the energy required to ionize one mole of the substance, and is usually stated in kilojoules per mole (KJ $mole^{-1}$). In chemistry, the terms 'second,' 'third,' etc., ionization potentials are usually used for the formation of doubly, triply, etc., charged ions.

Iron It is presented in 10 to 1000 parts per million parts in plant tissue. It is important in formation of chlorophyll in the plants, it is made available to the plants in Fe^{++} and Fe^{+++} forms. In the deficiency of this element, the stems of plants become thin and small. Leaves have spots of yellow and brown colour but the vein of the leaves remain green in acute deficiency, the whole leaf turn white. Iron is necessary for the synthesis of chlorophyll though it actually does not enter into its composition. Iron-starved plants develop chlorosis in the young leaves, the veins remaining green.

It is a transition element occurring in many ores, especially the oxides (hematite and magnetite) and carbonate. It is extracted in a blast furnace using coke, limestone, and hot air. The coke and air form carbon monoxide, which then reduces the iron ore to iron. The limestone removes acidic impurities and forms a layer of slag above the molten iron at the base of the furnace.

Steel is formed by reducing the carbon content to between 0.1 and 1.5 per cent. Iron is used as a catalyst in the Haber process for ammonia production. It corrodes in air and moisture to hydrated iron (III) oxide (rust). The most stable oxidation is +3m which is yellow, but +2 (green) and +6 easily reduced) also exist. Solution of iron (II) ions give a green precipitate with sodium hydroxide solution, whereas iron (III) ions give a brown precipitate. The concentration of iron (II) ions can be estimated in acid solution by titration with standard potassium manganate (VII) solution. Iron (III) ions must first be reduced to iron (II) ions with sulphur dioxide.

Symbol: Fe; m.p. 1539°C; b.p. 2887°C; r.d. 7.9; p.n. 26; r.a.m. 55.85.

Iron and soil pH When the pH of a mineral soil is low, appreciable amounts of iron, aluminium and manganese are active so much, so that the aluminium may become extremely toxic to certain plants.

Iron (III) oxide (ferric oxide, Fe_2O_3) A rusty-brown solid prepared by the action of heat on iron (III) hydroxide or iron (II) sulphate. It occurs in nature as the mineral hematite. Industrially it is obtained by roasting iron pyrites. Iron (III) oxide dissolves in dilute acids to produce solutions of iron (III) salts. It is stable at red heat, decomposes around 1300°C to give tri-iron tetroxide and can be reduced to iron hydrogen at 1000°C. Iron (III) oxide is not ionic in character but has a structure similar to that of aluminium (III) oxide.

Iron (II) sulphate (ferrous sulphate, green vitriol, $FeSO_4$ $7H_2O$) A compound that occurs in nature as the mineral melanterite (or coopers). It is made industrially from iron pyrites. In the laboratory iron (II) sulphate is prepared by dissolving excess iron in dilute sulphuric acid. On crystallization, green crystals of the heptahydrate are obtained. Careful heating of the hydrated salt yields anhydrous iron (II) sulphate; on further heating the sulphate decomposes to give iron (III) oxide, sulphur (IV) oxide and sulphur (VI) oxide. The hydrated crystals oxidize easily on exposure to air owing to the formation of basic iron (II) sulphate. A freshly prepared solution of iron (II) sulphate absorbs nitrogen (II) oxide (brown-ring test).

Iron (II) sulphate crystals are isomorphous with the sulphates of zinc, magnesium, nickel and cobalt.

Iron (III) sulphate (ferric sulphate, $Fe_2(SO_4)_3$ A compound prepared in the hydrated state by the oxidation of iron (II) sulphate dissolved in dilute sulphuric acid, using an oxidizing agent, such as hydrogen peroxide or concentrated nitric acid.

Ironpan A thin compact layer within the soil consisting of an accumulation of sesquioxides, notably of iron. In humid climates (*e.g.* in the British Isles) ironpans are characteristic of the B horizons of podzols and are formed by the redeposition of iron after leaching from above. *See* also hardpan.

Irreversible reaction A reaction in which conversion to products is complete; *i.e.* there is little or no back reaction.

Isocyanide Same as isonitrile.

Isocyanide test (carbylamine reaction) A test for the primary amine group. The suspected primary amine is warmed with trichloromethane in an alcoholic solution of potassium hydroxide. The resulting isocyanide (RNC) has a characteristic smell of bad onions and is very toxic:

$$CHCl_3 + KOH + RNH_2 - RNC + 3KCl + 3H_2O$$

Isoelectronic Describing compounds that have the same number of electrons. For example, carbon monoxide (CO) and nitrogen (N_2) are isoelectronic.

Isomer Anyone of the possible compounds arising from the different ways of grouping the atoms in a given molecular formula. For example, methoxymethane and ethanol are isomers of C_2H_6O. Types of isomers are classified according to the type of isomerism exhibited. *See* isomerism.

Isomerism The existence of two or more chemical compounds with the same molecular formula but different structural formulae or different spatial arrangements of atoms. The example, the compound C_4H_{10} may be butane (with a straight chain of carbon atoms) or 2-methyl propane (CH_3CHCH_3)CH_3; with a branched chain).

Structure isomerism is the type of isomerism in which the structural formulae of the compounds differ. There are two main types. In one the isomers are different types of compound. An example is ethanol (C_2H_5OH) and mehoxymethane (CH_3OCH_3), both having the formula C_2H_6O. In the other type of structural isomerism, the isomers differ because of the position of a functional group. For example, the primary alcohol propan-1-ol (C_3H_7OH) and the secondary alcohol propan-2-ol ($CH_3CH(OH)CH_3$) are isomers.

Isomorphism The existence of compounds with the same crystal structure.

Isomitrile (isocyanide) An organic compound of the formula R-NC.

Isotopes Two or more species of the same element differing in their mass numbers because of differing numbers of neutrons in their nuclei. The nuclei must have the same number of protons (an element is characterized by its proton number). Isotopes of elements are useful in chemistry of studies of the mechanisms of chemical actions.

Isotopic mass (isotopic weight) The mass number of a given isotope of an element.

Isotopic number The difference between the number of neutrons in an atom and the number of protons.

Isotopic weight *See* isotopic mass.

J

Jeweller's rough Iron (III) oxide (Fe_2O_3), used as a mild abrasive.

Joule symbol–J The SI unit of energy and work, equal to the work done when the point of application of a force of one newton moves one meter in the direction of action of the force. 1 J = 1 Nm. The joule is the unit of all forms of energy.

K

Kainit It is a mineral having mostly KCI and is used as fertilizer. Contains 12–20 per cent potash in addition to double this quantity of sodium chloride.

Kaolin Also known as China clay. A fine white clay that easily crumbles in powder, resulting from the alteration of feldspars in granite. The alteration is caused either by weathering (hydrolysis) or by the action of ascending gases associated with igneous activity. It is used in the manufacture of fine porcelain and china and as a filler in the manufacture of paper, rubber and paint.

Ketal A type of organic compound formed by addition of an alcohol to a ketone. Addition of one molecule gives a hemiketal; two molecules give a full ketal. *See* acetal.

Ketogenesis The formation of ketone bodies.

Ketohexose A ketose sugar with six carbon atoms. *See* sugar.

Ketone A type of organic compound with the general formula RCOR, having two alkyl or aryl groups bound to a carbonyl group. They are made by oxidizing secondary alcohols (Just as aldlehydes are made from primary alcohols). Simple examples are propanone (acetone, CH_3COH_3) and butanone (methylethyl ketone, $CH_3COC_2H_5$).

Ketone body One of a group of organic substances formed in fat metabolism, mainly in the liver. Examples are aceta-acetic acid and acetone. If the body has little or no carbohydrate as a respiratory substrate, ketosis occurs, in which more ketone bodies are produced than the body can use.

Ketomentose A ketose sugar with five carbon atoms.

Ketose A sugar containing a ketone (= CO) or potential ketone group.

Killed spirits Zinc (II) Chloride solution used as a flux for solder, so called because it is made by adding zinc to hydrochloric acid ('spirits of salt').

Kilo-symbol: K A prefix denoting 10^3. For example, 1 kilometre (km) = 10^3 metres (m).

Kilogram (kilogramme) Symbol: kg, The SI base unit of mass, equal to the mass of the international prototype of the k kilogram, which is a piece of platinum-iridium kept at Sevres in France.

Kilometre (kilometer) Symbol: km, An SI unit of length, equal to 1000 metres.

Kilowatt-hour Symbol: kWh, A unit of energy usually electrical, equal to the energy transferred by one kilowatt of power in one hour. It has a value of 3.6×10^6 joules.

Kinds of aridic soils Chernozems (black earths), Chestnut and reddish chestnuts, brown and reddish-brown chestnuts, grey desert (seirozems), reddish desert soils and non-calcic brown soils. There are formed in the arid or semiarid-temperature, semi-tropical their vegetation is grass and desert plants. Iron and aluminium do not have much movement or concentration in them. Calcium carbonate characteristically accumulated in the B horizon.

Kinds of humid soils These soils are in humid and semi-humid climates (cold, temperate, tropical), their vegetation is forests, and with tundra, mosses, lichens shrubs. Iron and aluminium have concentration and translocations in certain horizons and calcium carbonate have no accumulation in any mature horizon.

Kinetic theory A theory explaining physical properties in terms of the motion of particles. The kinetic theory of gases assumes that the molecules or atoms of a gas are in continuous random motion and the pressure (p) exerted on the walls of a containing vessel arises from the bombardment by these fast moving particles, when the temperature is raised the speeds increase; so consequently does the pressure. If more particles are introduced or the volume is reduced, there are more particles to bombard unit area of the walls and the pressure also increases. When a particle collides with the wall, it experiences a rate of charge of momentum, which is proportional to the force exerted. For a large number of particles, this provides a steady force per unit area (or pressure) on the wall.

Kinetic energy It is the energy possessed by a body due to its motion. If m-mass of body; V-its velocity; then K.E. = $1/2mV^2$.

Kinetic isotope effect *See* isotopes.

Kinetics A branch of physical chemistry concerned with the study of rates of chemical reactions and the effect of physical conditions that influence the rate of reaction, etc. The measurement of these rates under different conditions gives information on the mechanism of the reaction; *i.e.* on the sequence of processes by which reactants are converted into products.

Kipp's apparatus An apparatus for the production of a gas from the reaction of a liquid on a solid. It consists of three globes, the upper globe is the liquid reservoir. The middle globe contains the solid and also has a tap at which the gas may be drawn off. When the gas is drawn off, the liquid rises from the lower globe to enter the middle globe and reacts with the solid, thereby releasing more gas.

Koppen climate classification A climate classification first proposed in 1918 by the German climatologist W. Koppen and subsequently modified several times. In the classification the earth is divided into temperature zones based on mean monthly temperatures. The five major zones are:

1. Tropical, all months over 20°C.
2. Subtropical, 4–11 months between 10°C and 20°C.
3. Temperature, 4–12 months between 10° and 20°C.
4. Cold, 1 to 4 months between 10°C and 20°C and 8–11 months below 10°C.
5. Polar, all months below 10°C.

The temperature zones are further subdivided on the basis of seasonal incidence and amount of annual rainfall.

Krypton A colourless odourless monatomic element of the rare-gas group, now known to form compounds. It occurs in minute quantities (0.001 per cent by volume) in air. Krypton is used in fluorescent lights.

Symbol: kr; m.p. –157.3°Crs b.p. –153.4°C; d.3.73Kgm^{-3}; p.n. 36; r.a.m. 83.80.

L

Label A stable or radioactive nuclide used to investigate some process, such as a chemical reaction. *See* isotopes.

Lacolith (laccolite) A large dome-like mass of igneous rock that was intruded along a bedding plane in a sedimentary sequence of rocks, causing the overlying sediments to become arched above it.

Lactam A type of organic compound containing the –NH,CO.– group as part of a ring in the molecule. They can be regarded as formed from a straight-chain compound that has an amine group ($-NH_2$) at one end of the molecule and a carboxylic acid group (–COOH) at the other; *i.e.* from an amino acid. The reaction of the amine group, with the carboxylic acid group, with elimination of water, leads to the cyclic lactam, which is thus an internal amide.

Lacustrine Of or relating to a lake. For example, lacustrine sediments, which deposits are laid down within a lake. *See* also lacustrine delta.

Lacustrine delta A delta formed where a river slows into a lake, projecting from the shore of the lake out into the water. An example of this feature is founds in Switzerland where the river Rhone flows into Lake Geneva.

Lacustrine deposits *See* Glacial-lacustrine sediments, laevo-form. *See* optical activity.

Laevorotatory Describing compounds that rotate the plane of polarized light to the left (anticlockwise as viewed facing the oncoming light). Compare dextrorotatory. *See* optical activity.

Lamellar compound A compound with a crystal structure composed on thin plates or layers. Silicates form many compounds with distinct layers. Examples are talc ($Mg_3(OH)_2Si_4O_{10}$) and phrophylite ($Al_2(OH)_2Si_4O_{10}$). *See* also intercalation compound.

Land classification The grouping of land into different classes according to its physical characteristics. Various classification systems have been devised to measure land potential, *i.e.* its suitability for cultivation or other uses.

Land slides or slip erosion This is caused by hydraulic pressure exerted by heavy rains penetrating natural or man-made slopes too steep for stability in wet condition. On becoming saturated with water, great masses of soil and loose rock lying on steep slope comedown bodily. By this process, sometimes an entire field or hill side may side down.

Lapilli Fragments of cinder that are thrown out by a volcano during an eruption. The particles tend to be small, generally between 2 and 64 mm in diameter (Italian).

Lapse rate The rate of decrease of temperature with altitude expressed in °C per 100m, within the atmosphere as a whole, up to the level of the tropopause, the average decrease of temperature with altitude is about 0.6°C per 100m and this is known as the environmental lapse rate (ELR).

Latent heat The energy transferred when a substance changes state, *e.g.* from solid to liquid, liquid to gas, gas to liquid. In meteorology, the most important changes in state that occur are those from water vapour (evaporation) and from water vapour to water (condensation) within the atmosphere. In evaporation, the transfer of heat is from the atmosphere to the water, therefore the temperature of the atmosphere is reduced, while in condensation the transfer of heat is from the water vapour to the atmosphere therefore the temperature of the atmosphere is increased.

Lateral erosion The horizontal erosion carried out by a river on its banks. This acts chiefly on the outside (concave) bank of a meander bend where it causes undercutting of the bank and its eventual collapse. As a result each meander (and thus the valley) is widened and at the same time migrates slowly downstream. The higher land bordering the river is gradually eroded back leaving a low cliff known as a bluff bordering the meander belt.

Laterite A layer of deposits resulting from the deep weathering of rocks in humid tropical climates. The high temperatures and rainfall cause rapid removal of silica and alkalis from the weathered zone leaving a concentration of sesquioxides of iron and aluminium in the form of a thick red clay. The clay is plastic when wet but hardens once exposed to the atmosphere.

Laterite rocky mass The reddish honey-combed laterite rocky mass turns black after exposure to sun for sometime. The depth of the laterite murum layers varies considerably in different localities. On higher levels, these soils are exceedingly thin and gravely but on lower levels and in valleys, they are fairly deep and range from heavy loam to clays. Usually the laterite soils are poor in nitrogen, phosphoric acid, potash, lime and magnesia. While they are of low fertility on the whole, they respond readily to manuring and good cultivation. Valley soils can produce good crop of rice and sugarcane.

Laterite soils A red or yellow colour especially in the B horizon. The surface soil, if uneroded often is brown or grey. In some cases, red or yellow dominates the surface soil. This is the case when the parent material has been basaltic. The term laterite is derived from the Latin later, brick. It is now used to refer to natural materials characterized by an excess of sesquioxides and if well drained red in colour. The laterites are formed under conditions of high rainfall with alternating wet and dry periods. They are formed in situ by the leaching of bases and much of silica from the original rock. They are characterised by a compact to vesicular mass in the sub-soil horizons composed essentially of blocks being out while it is in the plastic state and then dried.

Laterization A weathering process in which iron and aluminium oxides become concentrated in the upper layers of the soil. This process takes place in tropical and subtropical regions with abundant rainfall and a marked dry season. Rocks, that are low in iron minerals but with a high proportion of aluminium, are converted by laterization to form bauxite, aluminium and manganese retarded. If drainage is good, intense oxidation occurs.

Lattice energy The energy released when ions of opposite charge are brought together from infinity to form one mole of a given crystal. The lattice energy is a measure of the stability of a solid ionic substance with respect to ions in the gas. *See* also Born-Haber cycle.

Laughing gas *See* dinitrogen monoxide.

Lava Molten or partially molten magma that is extruded from a volcano or a volcanic fissure out onto the surface of earth, where it cools and solidifies.

Law of conservation of energy *See* conservation of energy, law of.

Law of conservation of mass *See* conservation of mass, law of.

Law of constant composition *See* constant composition, law of.

Law of constant proportions *See* constant proportions, law of.

Law of isomorphism *See* Mitscherlich's law.

Law of mass action At constant temperature, the rate of a chemical reaction is directly proportional to the active mass of the reactants, the active mass being taken as the concentration in mol dm^{-3}. In general, the rate of reaction proceeds; a measure of the concentration of any of the reactants will give a measure of the rate of reaction. For the reaction A + B-products, the law of mass action states that:

$$\text{rate} = K(A)(B)$$

where, (A) represents the concentration of A in mol dm^{-3}. The term active mass can only equal the concentration in mol dm^{-3}, if there is no interaction or interference between the reacting molecules.

Laws of chemical combination *See* chemical combination, law of.

Law of diminishing returns An economic law stating that at a certain level of production, additional units of input will yield proportionately smaller units of output and the additional cost incurred will be greater than the additional revenue received. For example, the addition of more fertilizer to a field of wheat will at first increase the yield per hectare considerably, but a point will be reached at which the additional yield will have a lower value than the cost of the additional fertilizer which is applied.

Law of gravitation Everybody in this universe attracts every other body with a force which is directly proportional to the product of their masses and inversely proportional to the product of their masses and inversely proportional to the square of the distance between them; *e.g.* if m_1 and m_2 are the masses of two bodies, this distances between them. F is force of attraction between them, G is gravitational constant, then $F = G. m_1/m_2$. If $m_1 = m_2 = 1$ and d =1 then F = G.

Leaching It is the process by which water percolates downward due to gravitational movement of water in the soil. By this water movement, much of the portion of plant nutrients are drained with water. It is measured by the following flow of water:

A. Very small	0.05 inch per hour
B. Small	0.05–0.325 inch per hour
C. On average small	0.25–0.80 inch per hour
D. On average much	0.80–2.50 inch per hour
E. Much	2.50–5.00 inch per hour
F. Very high more than	5 inch per hour

Leaching losses of nutrients Due to variable weather conditions, to inherent differences in soils, and diversity in management especially to fertilization and cropping, it difficult to what the magnitude of the percolation losses of nutrients, from the soil is like to be.

Lead 1. A bluish-grey metal that is rarely found in its native state but most commonly occurs as the sulphide ore galena. This ore occurs in veins in igneous rocks or as the result of metamorphism.

2. A dense dull grey soft metallic element; the firth member of group IV of the periodic table. It occurs in small quantities in a wide variety of minerals but only a few are economically important. The most important one is galena (PbS), found in Australia, Mexico, the USA, and Canada. Other minerals are anglesite ($PbSO_4$), litharge (PbO) and cerussite ($PbCO_3$).

Galena is often associated with zinc ores and the smelting operations of both lead and zinc industries are closely integrated. The ore is concentrated by froth flotation and the concentrate roasted then reduced.

Like tin, forms halides PbX_2, with all halogens. $PbCl_4$ is the only known lead (IV) halide (all halides of SnX_4 are known). The great stability of Pb(II) state leads to Pb(IV) compounds being powerful oxidizing agents.

Symbol: Pb; m.p. 328°C; b.p. 1755°C; r.d. 11.34; p.n. 82; r.a.m. 207.19.

Lead arsenate A strong stomach insecticide.

Lead (II) carbonate-($PbCO_3$) A white poisonous powder that occurs naturally as the mineral cerussite. If forms rhombic crystals and can be precipitated by reacting a cold aqueous solution of a soluble lead salt (*e.g.* lead (II) nitrate) with ammonium carbonate.

Lead chamber process A former process for the manufacture of sulphuric acid by oxidizing sulphur (IV) oxide with nitrogen monoxide. The reaction, which was carried out in large lead chambers, has now been replaced by the contact process.

Least count The value of the division, on the circular scale (C. D.) of a screw gauge is called the least count, which is denoted as:

$$\text{Least Count} = \frac{\text{Pitch}}{\text{No. of divisions on circular scale}}$$

Leclanche cell A primary voltaic cell consisting, in its 'wet' form, of a carbon rod anode and a zinc cathode, with a 10–20 per cent solution of ammonium chloride as electrolyte. Manganese (IV) oxide mixed with crushed carbon in a porous bag or pot surrounding the anode acts as a depolarizing agent. The dry form (dry cell) is widely used for torch batteries, transistor radios, etc. It has a mixture of ammonium chloride, zinc chloride, flour and gum forming an electrolyte paste. Sometimes the dry cell is arranged in layers to form a rectangular battery, which has a longer life than the cylinder type.

Lee (leeward) The side away from the direction in which the wind is blowing, for example the side of a hill or mountain or a stretch of water sheltered from the wind. Compare windward.

Liebin condenser A simple type of laboratory condenser. It consists of a straight glass tube, in which the vapour is condensed, with a surrounding glass jacket through which cooling water flows.

Light reactions The light-dependent reactions of photosynthesis that convert light energy into the chemical energy of NASPH and ATP. *See* photosynthesis.

Lignite A type of coal that contains a high proportion of water and oxygen. It is brown or brownish-black in colour and crumbles when it is exposed to air long enough to dry out. It is formed mainly of wood, bark and leaves.

Lime *See* calcium oxide, calcium hydroxide.

Lime for soil fertility The influence of successively liming a soil over a period of years may tend to raise or lower its fertility according to system of management that accompanies the application of lime.

Lime present in soils In a humid region soil, it is present in about 0.60 per cent while in brown it is about 1.50 per cent.

Limestone Limestone contains carbonate of calcium. With magnesium carbonate, it occurs in dolomite. It is used in making calcium compounds, carbon dioxide and cement.

Limestone and oyster shell Serve as calcium supplements and grit in poultry rations.

Lime water A solution of calcium hydroxide in water. If carbon dioxide is bubbled through lime water, a milky precipitate of calcium carbonate forms. Prolonged bubbling of carbon dioxide turns the solution clear again as a result of the formation of calcium hydrogen-carbonate $(CaHCO_3)_2$.

Liming *See* flocculation.

Liming of soil It not only corrects acidity, but it also improves response to fertilizers, promotes desirable bacterial activity, improves soil structure and facilitates the growth of green manure and leguminous cover crops.

Limits of soil acidity Soil acidity exceeding a particular limit is injurious to plant growth. The availability of certain plant nutrients, particularly phosphorus, calcium, magnesium, iron and manganese becomes low in acid soils. Similarly, most of the desirable soil microbiological processes, such as the beneficial activities of Azotobacter and nodule bacterial of many legumes, are adversely affected. Satisfactory granulation of the soil also becomes difficult for achievement.

Limonite It is soil mineral of secondary origin, chemically it is $2Fe_2O_3.3H_2O$.

Limonite group Composition of hematite and of limonite group indicate the possibility of progressive change from red to yellow by hydration. This group contains gurgite, goethite, limonite, xanthodierite and limit.

Linoleic acid (linoic acid) A common unsaturated fatty acid occurring as glycerides in linseed oil, cottonseed oil and other vegetable oils.

Lipolysis The splitting of he component fatty acids from a lipid; *i.e.* part of the processes of catabolism of lipid molecules. Lipolysis is affected in the body, largely in the gut, by the lipase enzymes.

Lipopolysaccharide A conjugate polysaccharide in which the non-carbohydrate part is a lipid. Lipopolysaccharides are a constituent of the cell walls of certain bacteria.

Lipoprotein A conjugated protein formed by the combination of a protein with a lipid. Lipid material in mammalian blood is associated with plasma proteins, forming lipoproteins. Some lipoproteins are used to transport fats to metabolic sites whilst others are used to form the structures of cell membranes. They are found in mitochondria, nuclei and microsomes.

Liquid The state of matter in which the particles of a substance are loosely bound by intermolecular forces. The weakness of these forces permits movement of the particles and consequently liquids can change their shape within a fixed volume.

Liquid air A pale blue liquid. It is a mixture of liquid oxygen (boiling at –182.9°C) and liquid nitrogen (boiling at –195.7°C).

Lithium A light silvery moderately reactive metal; the first member of the alkali metals (group 1 of the periodic table). It occurs in a number of complex silicates, such as spodumene, lepidolite, and petalite, and a mixed phosphate, tryphilite. It is a rate element accounting for only 0.0065 per cent of the earth's crust. Lithium ores are treated with concentrated sulphuric acid and lithium sulphate subsequently separated by crystallization. The element is obtained by conversation to the chloride and electrolysis of the fused chloride or of solutions of LiCl in pyridine.

Lithium also forms a wide range of alkyl and aryl compounds with organic compounds, which are particularly useful in organic synthesis.

Lithium compounds impart a characteristic purple colour to flames.

Symbol: Li; m.p. 179°C, b.p. 1336°C; r.d. 0.82; p.n. 3; r.a.m. 6.9

Lithology The study of the physical and chemical characteristics of rocks, including colour, grain size, composition and texture. It is sometimes restricted to sedimentary rocks.

Lithosol A shallow stony azonal soil consisting chiefly of unweathered or partly weathered rock fragments, *e.g.* scree. Little soil development takes place and organic matter accumulates in the soil horizon giving it a dark appearance. Compare regosol.

Litmus A natural pigment that changes colour when in contact with acids and alkalis; above a pH of 8.3 it is blue and below a pH of 4.5 it is red. Thus it gives a rough indication of the acidity of basicity of a solution; because of its rather broad range.

Litmus paper Red or blue acid-alkali indicator papers. In acids, blue litmus paper turns red and in alkalis red litmus paper turns blue. The paper is prepared by soaking absorbent paper in litmus solution and then drying it. Litmus solution has a pH range from 4.5–8.3. It is obtained from a lichen (*Lecanora tartarea*).

Lixiviation The process of separating soluble components from a mixture by washing them out with water.

Loam A medium textured soil containing a mixture of large and small mineral particles. Loams are easy soils to work and combine the good properties of sandy and clay soils. This soil type, in mechanical composition is more or less midway between sand and clay. A loam may be defined as such a mixture of sand, silt and clay particles as to exhibit light and heavy properties in about equal proportions.

Loess All the loessial deposits are related to glaciation except that it was wind blown later when aridic conditions became prevalent. Loess is generally silty and has a yellowish-buff colour, unless it is very markedly weathered or carries any large amount of hums.

Loss by drainage of various soil constituents The amount of lime carried by drainage is always large for humid region soils, no matter what the soil condition or treatment may be. This accounts for the such soils becoming acidic in due course. There is a negligible loss of phosphorus by leaching due to the small amount present in the soil and the tenacity with which it is held in the soil. Obviously the losses of calcium, potassium and nitrogen by leaching are large wherever percolation is appreciable.

Loss of lime from soils These removals are large from humid soils, but also exert a drastic physiological effect on the soil. Lime is lost from the soil in three ways (1) by erosion (2) by crops and (3) by leaching.

Loss of nutrients due to erosion Erosion losses, even on a 4 per cent slope, may easily exceed the removal of nutrients by crops occupying the soil. Use of sound soil management principles, care in selecting the rotation sequence and the judicious use of sod crops are a matter of course.

Loss of nutrients in drainage Crops markedly reduce the removal of nutrients in drainage. This is especially true in respect to the nitrogen, the loss of which is very low from the cropped area. It is obviously poor management to allow a soil to remain barren during the summer in a humid region. Experiments indicated that nitrogen and phosphorus particularly with soil of medium to heavy texture are removed to a greater extent by crops, but the reverse is true in respect to calcium magnesium and sulphur. Potash is lost to drainage to the same degree as by crops.

Loss of nutrients through crops There is a negligible loss of phosphorus by leaching due to the small amount present in the soil and the tenacity with which it is held. Crops marked reduce the removal of nutrients through drainage, this being true for nitrogen, the loss of which is very low from cropped area.

Loss of nutrients by volatilization Nitrites in a slightly acid solution will evolve gaseous nitrogen when brought in contact with certain ammonium salts, with simple amines such as urea and even ammonium-nitrogenous sulphur compounds and carbohydrates. It should be noted that this type of volatilisation does not necessarily require adverse conditions such as poor drainage and aeration and may take place as commonly as does aminization, ammonification, and the oxidation of nitrites to nitrates.

Loss of organic matter It is well known that loss of organic matter through leaching is more in sandy soils than in soils having good amount of organic matter. Also nitrogen and potash are more susceptible to losses due to leaching than phosphorus. This loss cannot be made up in the soil without actual addition of these nutrients in the soil.

Luciferin A complex compound responsible for emitting light in bioluminescence.

Lumen Symbol–lm, The SI unit of luminous flux, equal to the luminous flux emitted by a point source of one candle in a solid angle of one steradian. 1 lm = 1 cd sr.

Luminescence The emission of radiation from a substance in which the particles have absorbed energy and gone into excited states. They then return to lower energy states with the emission of electromagnetic radiation. If the luminescence persists after the source of excitation is removed, it is called phosphorescence; if not, it is called fluorescence.

Lysimeter method This is actually the usual mode of approaching such problems as (1) In one type of lysimeter a small column of soil is entirely isolated by appropriate means from the land surrounding it. Effective and thorough drainage is provided. The advantage of this method is that the variations in a large field are avoided, the work of conducting the study is not so great, and the experiment is more easily controlled. (2) With other types of lysimeters, tanks, cans, or jars are used, the soil being supplied with adequate drainage. Experiments with such a method should be conducted for a considerable period in order to assume that the soil has been taken under more or less natural conditions.

M

Macroanimals of soil The larger animals of soils are–(1) rodents and insectivore, (2) insects, (3) millipedes, (4) wood lice, (5) mites, (6) slugs and snails, (7) centipedes, (8) spiders, and (9) earthworms.

Macroclimate The general climate of a large area such as Mediterranean climate.

Magma Molten rock material charged with gases and at a very high temperature lying beneath the earth's surface, on solidification and crystallization it forms igneous rock. When magma is extruded out onto the surface of the earth, it forms lava; if it solidifies within the crust, it either forms hypabyssal or plutonic rocks.

Magnesia *See* magnesium oxide.

Magnesite (magnesium carbonate, $MgCO_3$) The most common ore of magnesium. It is usually white in colour but can be grey, yellow or brown. It resumes from either the weathering of magnesium-rich or the replacement of dolomite and limestone under the action of volcanic gases.

Magnesium An element essential for plant and animal growth. It is contained in the chlorophyll molecule and is thus essential for photosynthesis. In animals, it is found in bones and teeth. As magnesium carbonate, it is found in large quantities in the skeletons of certain marine organisms, and is found in smaller quantities in the muscles and nerves of higher animals. It is an essential cofactor for certain phosphor enzymes, *e.g.* phosphohydrolase and magnesium is an essential constituent of chlorophyll. It is also necessary for formation of oil. It is usually needed by plants in relatively small quantities. Hence, its deficiency in soil is experienced later than that of potassium. Visual symptoms of an inadequate supply is the yellowing of the older leaves. In maize, the leaves develop intravenal white stripes; in cotton they change to purplish red, veins remaining dark green; in soyabean, they turn yellowish; and in apple trees, brown patches (blotches) appear on the leaves. Acute deficiency of magnesium also causes premature defoliation. There is an interaction between calcium and magnesium, and potassium.

Magnesium carbonate ($MgCO_3$) A white solid that occurs naturally as the mineral magnesite and in association with calcium carbonate in dolomite. Magnesium carbonate is sparingly soluble in water but reacts with dilute acids to give the salt, carbon dioxide and water.

Magnesium chloride ($MgCl_2$) A solid that exists in a variety of hydrated forms, most commonly as the hexahydrate, $MgCl_2.6H_2O$. When heated, this hydrated form is hydrolysed by its water of crystallization, with the evolution of hydrogen chloride and the formation of the oxide:

$$MgCl_2 + H_2O \rightarrow MgO + 2HCl$$

Magnesium hydrogencarbonate [magnesium bicarbonate, $Mg(HCO_3)_2$]. The solid hydrogencarbonate is unknown at room temperature. The compound is produced in solution when water containing carbon dioxide dissolves magnesium carbonate:

$$MgCO_3 + H_2O + CO_2 - Mg(HCO_3)_2$$

It is a cause of temporary hardness in water.

Magnesium hydroxide ($Mg(OH)_2$) A white solid that dissolves sparingly in water to give an alkaline solution. It can be prepared by adding an alkali to a soluble magnesium compound. Magnesium hydroxide is used in medicine as in antacid.

Magnesium loss by leaching It is invariably lost from the soil by leaching and drainage.

Mangnesium oxide (magnesia, MgO) A white solid that occurs naturally as the mineral periclase. It can be prepared by heating magnesium in oxygen or by decomposition of magnesium hydroxide, carbonate or nitrate. Magnesium oxide is weakly basis because of the attraction of the oxide ions for protons from water molecules:

$$O_2^- + H_2O - 2OH$$

Magnesium oxide melts at 2800°C and is widely used as a refractory lining for metal glass and cement furnaces.

Magnesium sulphate ($MgSO_4$) A solid that occurs naturally in many salt deposits in combination with other minerals.

Magnesium transference If we visualize the formation and solution of magnesium bicarbonate by hydrolysis and carbonation in one case and by ionic exchange in other, as to activity, magnesium seems to rival potassium but is much less mobile than calcium.

Magnetic declination (magnetic variation) The angle between true north and magnetic north, measured on a horizontal compass on the earth's surface. This varies with time and with position on the earth's surface. In 1982 declination at Greenwich was 4 × 57'W, decreasing about 12x a year.

Magnetism The study of the nature and cause of magnetic force fields, and how different substances are affected by them. Magnetic fields are produced by moving charge on a large scale (as with a current in a coil, forming an electromagnet), or on the small scale of the moving charges in the atoms. It is generally assumed that the earth's magnetism and that of other planets, stars and galaxies have the same cause. Substances may be classified on the basis on how samples interact with fields. Different type of magnetic behaviour result from the type of atom.

Magnate (loadstone, Fe_3O_4) A black magnetic iron ore containing 72 per cent iron. It is found in small quantities in most igneous rocks. Workable deposits of magnetite are found in the Urals, in the USSR, in the USA and at Kiruna and Gallivare in N Sweden.

Malachite A green mineral consisting of copper (II) carbonate and hydroxide ($CuCO_3Cu(OH)_2$). It is used as an ore and a pigment.

Maleic acid *See* butenedioic acid.

Malthusianism The ideas on population of the British economist, Thomas Malthus (1766–1834), which were put forward in his "Essay on the Principle of Population" (1798). He argued that population growth is a function of growth of the food supply. If the food supply was increased then the population would also grow. He propounded the idea that population, if unchecked, tends to increase at a geometrical rate *i.e.* 1, 2, 4, 8,......., while food production increase only at an arithmetical rate *i.e.* 1, 2, 3, 4,......., over the same time interval. He also told that population would increase to the limits of the means of subsistence and is only checked by war, famine and pestilence. Population increase would result misery and lower standard of living.

Maltose ($C_{12}H_{22}O_{11}$) A sugar found in germinating cereal seeds. It is a disaccharide composed of two glucose units. Maltose is an important intermediate in the enzyme hydrolysis of starch. It is further hydrolysed to glucose.

Manganate (VII) (permanganate) A salt containing the ion MnO_{4x} manganate (VII) salts are purple, and strong oxidizing agents. In basic solutions the dark green manganate (VI) ion is formed.

Manganese A transition metal occurring naturally as oxides, *e.g.* pyrolusite (MnO_2). Its main use is in alloy steel made by adding pyrolusite to iron ore in an electrical furnace. Manganese decomposes cold water and dilute acids to give hydrogen and reacts with oxygen and nitrogen when heated. The oxidation states are +7, +6, +4 and (the most stable) +2 manganese (II) salts are pale pink and with alkali the solutions precipitate manganese (II) hydroxide, which rapictly oxidizes in air to brown manganese (III) oxide.

Symbol: Mn; m.p. 1244xC; b.p., 2040xC; r.d. 7.4; p.n. 25; r.a.m. 54.94. *See* also trace element.

Manganese deficiency Manganese deficiency leads to a chlorosis in the intervenial tissue of net-veined leaves and plants with leaves that have parallel veins develop a general chlorotic condition and secondary symptoms. Symptoms of manganese deficiency in cereals produce grey speck, white streak, dry spot and leaf spot; in field peas, marsh spot; sugarcane, streak disease and Pahala blight; and in spinach and beans it is yellow disease.

Mannitol A soluble sugar alcohol (carbohydrate) found widely in plants and forming a characteristic food reserve of the brown algae. It is a hexahydric alcohol, *i.e.* each of the six carbon atoms has an alcohol (hydroxyl) group attached.

Mannose A simple sugar found in many polysaccharides. It is an aldohexose, isomeric with glucose. Mannose ($C_6H_{12}O_6$).

Manometer A device for measuring pressure. A simple type is a U-shaped glass tube containing mercury or other liquid. The pressure difference between the arms of the tube is indicated by the difference in heights of the liquid.

Mantle That portion of the interior of the earth lying beneath the crust and above the core. Evidence obtained from the records of earthquakes' waves suggests that the mantle is made up of ultrabasic rock. Its upper limit is horovicic discontinuity, which lies at an average depth of 35 Kms below the surface of the earth, and it descends as far as the Guetenberg discontinuity, which lies at about 2900 Kms.

Marble A limestone that has been completely recrystallized by metamorphic processes involving pressure and heat. The resulting rock is hard, patterned and shiny, with no trace of its original fossil content.

Marginal land Land that is sufficiently fertile to yield a return that just covers the cost of production. Marginal lands are abandoned if the price obtained for the products falls or if the cost of production increases.

Marine soils Made of materials carried by rivers and deposited in the ocean. Marine deposits have been worn and triturated by a number of agencies. These are weathering, eroding and emerge above sea and face the final action of the natural forces of weathering.

Marime climate Also called oceanic climate. A type of climate in which the ocean has a dominant influence. It is therefore found on islands and near coasts especially those having prevailing on shore winds.

Market The place at which there is a demand for commodities.

Market gardening The intensive cultivation of vegetables, fruits or flowers for sale. This form of cultivation is highly specialised and involves considerable capital and/or labour expenditure per unit area.

Marl A soil containing a high proportion of calcium carbonate, as in limestone area. Marl 1. A type of soil that is a natural mixture of clay and calcium carbonate. It is sometimes used as a fertilizer and for making bricks. 2. A mudstone containing a high proportion of calcareous material. *See* also Bog lime.

Marsh A poorly drained low-lying tract of land that possesses distinctive flora. The marsh be subject to occasional flooding and water table is nearly always very close to the surface.

Marsh gas Methane produced is marshes by decomposing vegetation.

Mass action, law of, *See* law of mass action.

Mass-energy equation The equation $E = mc^2$, where E is the total energy (rest-mass energy + kinetic energy + potential energy) of a mass m, c being the velocity of light. The equation is a consequence of Einstein's special theory of relativity; mass is a form of energy and energy also has mass.

Mass spectrometer An instrument for producing ions in a gas and analysing them according to their charge/mass ratio. The earliest experiment by J.J. Thomson used a stream of positive ions from a discharge tube, which were deflected by parallel electric and magnetic fields at right angles to the beam. Each type of ion formed a parabolic trace on a photographic plate.

Matte A mixture of iron and copper sulphides obtained at an intermediate stage in smelting copper ores.

Maturity and old age of soils Material so also in desert area, soils are constantly subject to fresh deposition of wind-blown sand. In these situations, therefore, the soils may never reach maturity or old age.

Meadow soil An intrazonal hydromorphic soil that occurs in low-lying area such as floodplains. It has a humus rich dark brown colour. A horizon-that overlies a mottled gleyed zone. Lush grassland, often managed for a hay crop, provides the surface humus. The gleying results from a high water table or recurrent flooding. *See* also gley soil.

Mechanical advantage It is the ratio between the weight lifted (W) and the power applied (P). Thus M.A. = W/P.

Mechanical analysis In order that the mineral particles of a soil which vary so widely in size may be studies with success, they must be separated into convenient groups. The various groups are spoken as separates.

Mechanical soil reclamation The reclamation of saline soils by permanent lowering of water table to five or six feet below the surface. The provisions of a network of deep cuts at right angles to the slope of land serves the purpose. This should be followed by leaching suffice to remove all harmful salts. Leaching also drives down considerable quantities of salts below the ordinary root zone.

Mechanical soil conservation measures These measures are adopted to supplement the biological methods when the latter alone are not adequately effective. These measures should not only be technically sound but also economically feasible. The important principle involved in the design of all mechanical structures to control erosion are:

1. Increasing the time of concentration and thereby allowing more runoff water to be absorbed and held by soil.
2. Dividing a long slope into several short ones so as to maintain a minimum velocity for the runoff water.
3. Protection against damage due to runoff.

Mechanical weathering (physical weathering) The break-up of rock by processes of weathering that do not involve chemical change. Mechanical weathering is responsible for the disintegration for rocks, as opposed to decomposition, which results from chemical weathering. The physical stresses and strains causing mechanical weathering may be the result of expansion and contraction of grains or crystals in the rock itself or the result of changing external pressures. These processes include thermal expansion, which results in exfoliation; freeze thaw action/salt-crystal growth, in which the growth of crystals creates pressure in the rock; pressure release (unloading) in which the removal of surface rocks leads to the splitting of rocks; and animal and plant activity. *See* also exfoliation, freeze-thaw, granular disintegration.

Melting (fusion) The process by which a solid is converted into a liquid by heat or pressure.

Melting point The temperatures at which a solid is in equilibrium with its liquid phase at standard pressure and above which the solid melts. This temperature is always the same for a particular solid. Ironically bonded solids generally have much higher melting points than those in which the forces are covalent or intermolecular.

Mercuric chloride (Corrosive sublimate) In a solution of one in 1,000 parts of water, it is perhaps the strongest disinfectant known, but it corrodes metals and is an exceedingly poisonous substance. It should be used with great caution.

Mercuric oxide *See* mercury (II) oxide.

Mercurous sulphide *See* mercury (I) sulphide.

Mercury (quicksilver) A silver-white metal that is liquid at normal temperature and pressure. It is obtained chiefly from its ore cinnabar, though it is also found in the native state.

Mercury cell A voltaic or electrolytic cell in which one or both of the electrodes consists of mercury or an amalgam. Amalgam electrodes are used in the Daniell cell and the Weston cadmium cell.

Mesozoic The geological era that extended from about 225 million years ago. It followed the Palaeozoic era and preceded the Cainozoic era.

Mesquite A spiny shrub or small tree of the genus Prosopis that grows in the arid climate.

Metals Any of a class of chemical elements with certain characteristic properties. In everyday usage, metals are elements (and alloys) such as iron, aluminium and copper which are lustrous metaleable solids-usually good conductors of heat and electricity. Note that this is not a strict definition-some metals are poor conductors, mercury is a liquid.

In chemistry, metals are distinguished by their chemical properties and there are two main groups. Reactive metals such as the alkali metals and alkaline-earth metals are electropositive elements. They are high in the electrochemical series and tend to form compounds by losing electrons to give positive ions.

They have basic oxides and hydroxides. This typical metallic behaviour decreases across the periodic table and increases down a group in the table.

The other type of metals are the transition elements, which are less reactive, have variable valencies and tend to form complexes. In the solid and liquid states metals have metallic bonds, formed by positive ions with free electrons. *See* also metalloid, non-metal.

Metamorphic rock Rock formed when pre-existing sedimentary or igneous rock is altered as a result of changes in physical of chemical conditions. This process of metamorphism may be through intense stress or pressure caused by earth movements, increased temperature caused by volcanic activity, or the action of gases and liquids of magnetic origin.

Meteorology Study of the atmosphere and the physical processes at work within it including pressure; winds, temperature, precipitation, clouds, sunshine, etc.

Meteorological satellites Spare craft placed into orbit around the earth and equipped with instruments for recording meteorological data.

Methanol (formaldehyde, HCOH) A colourless gaseous aldehyde. It is manufactured by the oxidation of methanol (500xC and a silver catalyst) $2CH_3OH + O_2 - 2HCOH + 2H_2O$. The compound is used in the manufacture of urea formaldehyde resins.

Methyl orange An acid-base indicator that is red in solutions below a pH of 3 and yellow above a pH of 4.4. As the transition range is clearly on the acid side, methyl orange is suitable for the titration of an acid with a moderately weak base, such as sodium carbonate.

Metre symbol m The SI base unit of length, defined as the length equal to 1650763.73 wavelengths in vacuum corresponding to the transition between the levels 210 and 5ds of the krypton-86 atoms.

Metric system A system of units based on the metre and the kilogram and using multiples and submultiples of 10. SI units, c.g.s units, and m.k.s units are all scientific metric systems of units.

Mica Mica is silicate of aluminium and potassium, with silicates of iron, magnesium and sodium.

Micelle A particle carrying a negative charge is for convenience sake is known as micelle or acidiod or nucleus.

Micro Symbol, A prefix denoting 10^{-6}. For example, 1 micrometre (μm) = 10^{-6} metre (m).

Micro Small.

Microanimals of soil There is an abundance of microanimals life in the soil. Two important groups are the nematodes and protozoa. Yet another group is of rotifers.

Microbial cultures and their enzymes These are the basic elements in applied microbiology and biotechnological process as they represent the catalysts for these processes.

Microclimate The climate of the immediate surrounding of a particular object of study.

Microcline It is soil mineral of primary origin represented as $KalSi_3O_8$. It is a source of potash in the soil.

Millibar (mb) A unit of pressure used in meteorology. It is equal to a thousandth of a bar or 100 pascals.

Milliequivalent 1. One milliequivalent of any cation is equivalent on the basis of the acre furrow slice (2000000 lbs of dry soil) to 20 lbs of hydrogen 0 to 1000 lbs of $CaCO_3$; this is a convenient conversion.

2. In order that the amounts of such exchangeable cations as Ca, Mg, K etc. may be accurately compared with the amounts of adsorbed H_2, they must be reduced to milliequivalents. Suppose that 40 mgs of exchangeable Ca are present in 100 gms of dry soil. Since one atoms of Ca (atomic wt. 40) will replace 2 atoms of H (atomic wt. 1), the weight relation is 40 : 2 or 20 : 1. Hence the 40 mgs of Ca is equivalent to mgs of H or to 2 milliequivalent.

Millimetre (mm) = 10^{-3} metre (m).

Millimetre of mercury *See* mm Hg.

Millimicron *See* nanometre.

Million's reagent A mixture of mercury, nitric acid and water that is used in testing for the presence of proteins. The reagent is heated with the test material and the production of brick red precipitate indicates the presence of protein.

Mineral A naturally occurring substance having a definite chemical composition and atomic structure, and formed by inorganic process. About 2000 minerals are known. Rocks are made up of various combinations of minerals. Some minerals form a series in which there is gradual replacement of one element by another, *e.g.* the feldspars. The most common mineral in rocks is quartz. The term is also used for any naturally occurring material that is mined and of economic value.

Mineral composition of soil Due to a varying speed of the wearing down of different minerals, the more easily decomposing feldspars predominate in finer separates, whereas the resistant minerals like quartz are more abundant amongst the coarse particles. Due to this, the separate show differences in chemical composition also.

Mineral manures Such manures which are prepared from mineral resources. Most of the chemical fertilizers are in this class.

Mineral matter and its composition The principal minerals occurring in the earth's crust are the following:

Feldspar	48 per cent
Quartz	36 per cent
Mica	10 per cent
Limestone and Magnesium limestone	2 per cent
Hornblende and Augite	1 per cent
Olivine and Serpentine	1 per cent
Clay	1 per cent
Other minerals	1 per cent

Mineralogy the scientific study of minerals.

Mineral oil 1. These are effective insecticides but if used undiluted, they will severely damage foliage and even wood.

2. A dry mineral oil absorb about one-fifth to two-fifth of water of its weight, a organic soil will absorb more due to its colloidal nature about 2–3 times of its dry weight. The water bonding capacity is much less.

Mineral spring A spring of water that contains a high proportion of dissolved mineral salts such as iron compounds, magnesium chloride and sodium chloride. Medicinal properties have frequently been attributed to such waters.

Mineral versus organic soils A mineral soil may be high or low in organic matter. Soils low in organic matter are referred to as mineral soils since the inorganic constituents separately predominate. Most soils are of this nature. But in swamps, bogs and marshes, conditions are often such as to encourage the accumulation of much organic matter. This gives rise to peat (organic) deposits and peat soils in which the organic content may range as high as 95 per cent.

Miocene The geological epoch of the tertiary period following the oligocene epoch which ended about 22.5 million years ago and preceding the pliocene epoch.

Mitscherlich's law (law of isomorphism) The law stating that substances that crystallize in isomorphous forms (*i.e.* have identical crystalline forms and form mixed crystals) have similar chemical compositions. The law can be used to indicate the formulae of compounds. For instance, the fact that chromium dioxide is isomorphous with Fe_2O_3 and Al_2O_3 implies that is formula is Cr_2O_3.

Mixed farming Farming that includes the growing of arable crops and the keeping of livestock. In many cases part of the arable and is used for growing fodder crop for the livestocks.

Mixture (1) The chemical properties of the components of a mixture are the same as those of the pure substances.

(2) The mixture can be separated by physical means (*e.g.* distillation of crystallization) or mechanically.

(3) The proportions of the components can vary. Some mixtures (*e.g.* certain solutions) can only vary in proportions between definite limits.

Modern detailed soil survey It is a comprehensive map and text portrayal of the kinds, amounts and geographical distribution of individual soils. It also gives information needed for planning land.

Moisture equivalent It is an arbitrary point of the pF curve obtained by subjecting a sample of wet soil, held in a vessel of specific dimensions, to a centrifugal force of 1000 times of gravity. Due to the condition to which the short column of soil is subjected to the pressure is applied, the average tension is equivalent to a pull on free column of only 500 times the force of gravity. Hence the moisture equivalent corresponds to a pF of 2.7 or about ½ atmosphere.

Molar Denoting a physical quantity divided by the amount of substance will be moles. For examples, volume (V) divided by the number of moles (n) is molar volume Vm = v/n. Molar solution contains one mole of solute per cubic decimetre of solvent.

Molecular crystal A crystal in which molecules, as opposed to atoms, occupy lattice points. Examples include iodine and solid carbon dioxide (dry ice). Because the forces holding the molecules together are weak, molecular crystals have low melting points. When the molecules are small, the crystal structure approximates to a close-packed arrangement. *See* close packing.

Mole drain It is an unwalled cylindrical channel, 3 to 4 inches in diameter, located below the soil surface and with a grade sufficient to provide for the satisfactory flow of water.

Molecular formula The formula of a compound showing the number and types of the atoms present in a molecule, but not the arrangement of the atoms. For example C_2H_6O represents the molecular formula both of ethanol (C_2H_5OH) and methoxymethane (CH_3OCH_3). The molecular formula can only be determined if the molecular weight is known. Compare empirical formula, structural formula.

Molecule weight *See* relative molecular mass.

Molecule A particle formed by the combination of atoms in a whole-number ration. A molecule of an element (combining atoms are the same, *e.g.* O_2) or of a compound (combining atoms are different, *e.g.* HCl) retains the properties of that element or compound. Thus, any quantity of a compound is a collection of many identical molecules. Molecular sizes are characteristically 10^{-10} to 10^{9} m. Many molecules of natural products are so large that they are regarded as giant molecules (macromolecules); they may contain thousands of atoms and have complex structural formulae that require very advanced techniques of identify. *See* also formula, relative molecular mass.

Mole Symbol: mol The Sl base unit of amount of substance that contains as many elementary entities as there are atoms in 0.012 kilogram of carbon-12. The elementary entities may be atoms, molecules, ions, electrons, photons, etc., and they must be specified. The amount of substance is proportionality being the Avogadro number. One mole contains 6.022045×10^{23} entities. One mole of an element with relative atomic mass A has a mass of A grams (this mass was formerly called one gram-atom of the element). One mole of a compound with relative molecular mass M has a mass of M grams (this was formerly called one gram molecule of the compound.

Molishch's test *See* alpha-naphthol test.

Molybedenum In absence of this element, the leaves show mottling and leaves start curving of the lamina and the margins start breaking. It is present in plant tissue in very small quantity 0.01 to 10 ppm, plant take in the form of Molybdate MoO_4. Molybedenum deficiency in Cruciferae-Causes extreme leaf malformation and unsatisfactory curd (head) formation. Apply lime to acid soils, and ammonium molybedate @ 1-16 lbs/acre.

Moment This of the force about any point is the turning tendency of the force about the point. It is measured as the product of the force applied and the length of perpendicular from the point on the line of action of force. If F is the force applied, d is the perpendicular distance (ON), and M is the moment of force; then $kM = F \times d$. The moment is taken positive if it tends to rotate the body in the anticlockwise direction and it is taken as negative if the rotation is clockwise.

Momentum It is the quantity of motion possessed by a body, and is measured by the product of its mass and velocity.

Monosaccharide A sugar that cannot be hydrolysed to simpler carbohydrates of small carbon content. Glucose and fructose are the examples.

Monotropy The existence of a single allotrope of an element that is always more stable than the other(s) regardless of temperature; phosphorus, for example, exhibits monotropy. The phase diagram for a monotropic element shows that one allotrope always has a lower vapour pressure that the other(s) at all temperatures; this is the stable allotrope.

Sl.No.	Morphological Description of Soil for Test	
1.	Colour	(*i*) Field
		(*ii*) Air dry
		(*iii*) Moist
2.	Texture	(*i*) Field
		(*ii*) Air dry
		(*iii*) Moist
3.	Structure	
	(*a*)	(*i*) Field
		(*ii*) Air dry
		(*iii*) Moist
	(*b*) Cementation	
	(*c*) Concretions	
4.	Organic matter	
5.	Roots (feeders)	
6.	Consistency	
7.	Remarkable peculiarities of special features, if any.	

Monsoon peats The pea soils are generally submerged under water during the monsoon season. As soon as the rain ceases they are put under paddy cultivation. The lands are black, heavy and highly acidic, pH sometimes being as low as 3.9. They contain 10 to 40 per cent organic matter and 20–30 per cent of Fe_2O_3 and AbO_3. They are comparatively poor in available P_2O_5. Investigations on these soils is due to the decomposition of organic matter under anaerobic conditions and no nitrification is at all possible, and sometimes the soils are highly toxic to plant life as they contain ferrous and aluminium sulphates in considerable amounts.

Montmorilonite It is composed molecularly of two silica sheets and one of alumina. Of the three common of this group, montomorilonite, beidellite and nontronite, the first two are found in the soils in large amounts. The molecules of these clays are found in the soils firmly linked together than those of kaolin group and therefore are usually farther apart.

Mountain and hill soil The study on the nature of forest soils is essential, particularly in connection with any project of a forestation. The formation of these soils is mainly governed by the characteristic deposition of the organic matter derived from the forest growth.

Movement of soil water *See* soil water movement.

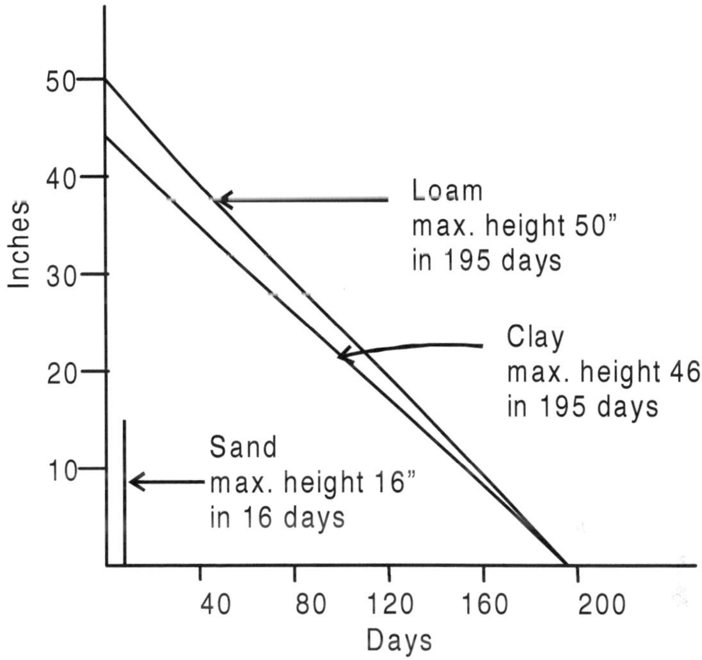

Upward moisture movement from a water-table through soils of different texture and structure of soils

Movement of humus in soil The movement and distribution of humus in the soil is either mechanical or in suspensions. Normally, the amount of organic matter, and particularly humus, decreases from the surface downward until it disappears completely. This downward decrease is sudden in woodland soils and very gradual in grasslands. In forests, the falling leaves add considerable quantities of organic matter to the surface soil. If soil acidity increases, humus becomes dispersed and is carried down into the sub-soil where it may form a part of a hard layer as peat.

Soil Group	*Percentage Content of Humus*
Black cotton soil	0.34 to 0.77
Red soil and laterites	0.68 to 1.53
Alluvial soils	0.28 to 1.10

Muck soils Organic deposits in soils are classified on the basis of the amount of organic matter that they contain. When the quantity of this constituent ranges for 20 to 50 per cent, such a soil is called muck.

Mud A fine-grained elastic sediment in which most of the particles measure less than 0.06 mm. It consists of particles of minerals produced by both physical and chemical weathering. Muds are deposited on lake beds, deep-sea bottoms, and close to shore, where they are mud flats. In the geological record, they are represented by mud-stones.

Mull A form of humus with a pH value in the range of 4.5–6.5 that develops under grassland and some broad-leaved forests in well aerated and well drained soils, such as brown earths and rendzinas. Mull is produced by the relatively rapid decomposition of nutrient rich plant litter. The presence of soil fauna, especially, earthworms, aids the rapid integration of the humus into the soil. Mull is characteristically well integrated with mineral matter, rich in proteins, basis of nitrogen, and conducive to curb structure. Compare mor, *See* humus. When the organic matter becomes more and more mixed with the mineral soil below until A_0 is very thin, most of the organic matter being a part of A_1. This layer now called mull, thus changing from a grey to brown colour.

Multiple proportions, law of, proposed by Dalton in 1804, the principle that when two elements A and B combine to form more than one compound, the weights of B that combine with a fixed weight of A are in small whole-number ratios, for example, in nitrous oxide, N_2O, nitric oxide, NO, and dinitrogen tetroxide, N_2O_4, the amounts of nitrogen combined with a fixed weight of oxygen are the ratio 4 : 2 : 1.

Multiple-range indicator *See* universal indicator, mull layer: Certain forest and grassland soils having organic matter accumulations due to earth worms in the A horizon called mull layer.

Mycorrhiza of soil The mycelia of certain members of fungi often infest the roots of trees giving an association called mycorrhiza. This apparently is advantageous both to the host and the parasite.

Mythyl red An acid-base indicator that is red in solutions below a pH of 4.2 and yellow above a pH of 6.3. It is often used for the same types of titration as methyl orange but the transition range of methyl red is nearer to neutral pH7 than of methyl orange. The molecules are structurally similar.

N

Naming of soils Much can be judged as to the texture and class of soil merely by rubbing it between the thumb and the fingers or in the palm of the hand as by any other superficial means. *See* also mechanical analysis.

Naphthalene ($C_{10}H_8$) A white crystalline solid with a distinctive smell of mothballs. Naphthalene is found in both the middle and heavy-oil fractions of crude oil and is obtained by fractional crystallization. It is used in the manufacture of benzene-1.2-dicarboxylic anhydride and thence in the production of plastics and dyes.

Natural gas *See* gas.

Natural soil building The removal of the surface soil from areas denuded of natural protective cover as a result of human and animal interference takes place at a much faster rate than it is built up by the soil forming processes. This accelerated detachment rapidly ravages the land and it is with this type of soil erosion that we are so seriously concerned. Nature requires normally 1,000 to 2,000 years to build an inch of top soil but wrong farming methods may take only a few years to erode the same from lands of average slopes.

Nature of amino acids Amino acids are produced by replacing one of the alkyl hydrogen with NH_3. Acetic acid (CH_3COOH) thereby becomes amino-acetic acid or glycol (CH_3NH_2COOH). Amides or acid amides are formed from organic acids by replacing the hydroxyl becomes acetamide (CH_3CONH_2).

Nature of aridic soils It includes all soils whose profile characteristics are adjusted to a low rainfall, usually not over 25 inches annually. The temperature may be low, medium or high. A fully and normally developed zonal profile usually reveals a calcium carbonate accumulation at some point in the A and B horizons higher in percentage than the lime content of the parent material. As the rainfall decreases this layer of calcium carbonate, accompanied by more soluble salts, is found nearer the surface.

Nature of ground cover When rain falls on a surface covered by a thick mantle of plants, its velocity and the erosive powers are reduced and most of the water either quickly percolates into the soil or moves over the surface with non-erosive velocity. Areas not protected with thick cover of plants are unable to absorb water effectively because the dashing rains shatter the soil surface, the fine soil particles go into solution or suspension and the thick mixture of water and soil quickly fills and closes the tiny interstices in the soil, reducing infiltration and consequently increasing run-off and soil loss.

Nature reserve An area designated for the preservation of wild life and plants safeguards a single rare species but more usually positive management of the area.

Need for soil and water conservation The need for soil and water conservation is, therefore, very vital and urgent for supporting the rapidly growing population of India at a reasonable standard of living. In serious of the problem of food production is apparent from the fact that in many years food-grains have to be imported from abroad in substantial quantity.

Need for soil testing It is well known that one of the chief factors, responsible for satisfactory crop yields, is the presence of essential plant nutrients in soil in adequate quantities and in readily utilisable form. For maximum production and intelligent soil management, a knowledge of the fertility status and physical properties of a soil is essential. In fact, soil testing forms an essential part of any scheme of scientific development of land. Soil testing facilities are now provided by all State Agricultural Departments as well as the Indian Agricultural Research Institute, Delhi.

Nematodes or round worms Nematodes are elongated, cylindrical and thread-like worms, sexually differentiated into males and females. Most nematode species lay numerous eggs containing embryo larvae in different stages of development, but some give birth to fully developed larvae. The process whereby the eggs and larvae develop outside the animal body, eventually growing into the effective stage, varies within wide limits, but the adult stage is reached inside the animal body. Some species require 'intermediate hosts' for a part of their development cycle.

Nematodes in soil Nematodes, thread-worms and eel-worms as they are called are found in almost all kinds of soils, often in large numbers. A maximum of 50 to the gram of dried soil would mean about 45 billions to the acre.

Neon An inert colourless odourless monatomic element of the rare-gas group. It occurs in minute quantities (0.0018 per cent by volume) in air. It is used mainly in neon signs and lights. Symbol Ne; m.p. –248.67°; b.p. –246.05°; d. 0.9 kgm^{-3}; p.n. 10; r.a.m. 20.18.

Neoprene A type of synthetic rubber made by polymerization of 2-chlorobuta-1,2,-diene ($H_2C:CHCCl:CH_2$). It is more resistant to oil and temperature than natural rubbers.

Neutralization The stoichiometric reaction of an acid and a base in volumetric analysis. The neutralization point or end point is detected with indicators.

Neutron number Symbol: N The number of neutrons in the nucleus of an atom; *i.e.* the nucleon number (A) minus the proton number (Z).

Newton Newton is equal to 1 kg × 1 m/sec^2; which is equal to 1000 gm × 100 cm/sec^2, which is equal to 10 gm × cm/sec^2 and is equal to 10^5 dynes.

Symbol: N, the SI unit of force, equal to the force needed to accelerate one kilogram by one meter $second^{-2}$, 1 N = 1 $kgms^{-2}$.

Nickel A metal that never occurs in its native state but is found in sulphide and arsenide ores, such as millerite and pentlandite. It is white, unaffected by exposure to air, and capable of being highly polished. It is used in coinage and in the manufacture of steel alloys, such as stainless steel, Inver and Monel. Because nickel steel is very hard, it is used in the construction of cars and aircraft. One of the main sources of nickel is the nickel-iron sulphide deposits near Sudbury, Ontario, which make Canada the world's primary nickel producer. The earth's core is believed to contain large quantities of nickel.

Nickel-silver (German silver) Any of a group of alloys containing nickel, copper and zinc in various proportions (but no silver). They are white or silvery in colour, can be highly polished and have good corrosion-resistance. They are used, for example, in silver plating chromium plating and enamelling.

Ninhydrin A colourless organic compound that gives a blue coloration with amino acids. It is used as a test for amino acids; in particular, to show the positions of spots of amino acids in paper chromatography.

Nitrate accumulation in organic soils There is no type of organic soil that can be used as a manure. Organic soils have a wide nitrogen-carbon ratio the minimum being in the neighbourhood of 1 : 20. Organic in spite of their high nitrogen ration show exceedingly, even in a low-lime organic soil, is usually far greater than that of a representative mineral soil.

As far as accumulation in green manured field is concerned, due to the addition of fresh decomposable plant tissues results in a rapid multiplication of organisms of all kinds. As a consequence, nitrates disappear from the soil immediately and no nitrogen in this form may accumulate until the period of rapid decay is past.

As decomposition of plant tissues go further, both carbon and nitrogen become subject to loss and it is only a question of time until their relative rate of disappearance from the soil becomes constant, 9 or 10 parts of carbon being oxidised to one part of nitrogen change to the nitrate.

Nitrate of soda Chile saltpetre ($NaNO_3$) It is soluble in water and on crystallization forms colourless deliquescent crystals. It is used as a fertilizer and as a source of nitrate and nitric acid.

Nitric acid (HNO_3) A colourless fuming corrosive liquid that is a strong acid, by the distillation of a mixture of an alkali metal nitrate and concentrated sulphuric acid. Commercially it is prepared by the catalytic oxidation of ammonia and is supplied as concentrated nitric acid, which contains 68 per cent of the acid and is often coloured yellow by dissolved oxides of nitrogen. Nitric acid is a strong oxidizing agent. Most metals are converted to their nitrates with the evolution of oxides of nitrogen (the composition of the mixture of the oxides depends on the temperature and on the concentration of the nitric acid used). Some non-metals (*e.g.* sulphur and phosphorus) react to produce oxyacids. Organic substances (*e.g.* sawdust and ethanol) react violently, but the more stable aromatic compounds, such as benzene and toluene, can be converted to nitro compounds in controllable reactions.

Nitrobacter The nitrifying organisms are collectively called nitrobacter. Nitrosomonas and Nitrosococcus are concerned in the conversion of ammonia into nitrites. The organism having to do with the oxidation of nitrites to nitrates are generally designated as nitrobacter.

Nitrogen Encourages the vegetative development of plants, importing a healthy green colour to the leaves. It also controls to some extent the efficiency of utilization of phosphorus and potassium. Its deficiency retards growth and root development, turns foliage yellowish or pale green, hastens maturity, causes shrivelling of grain and lowers crop yield. The older leaves are affected first. An excess if nitrogen produces leathery (and sometimes crinkled), dark green leaves and succulent growth. It also delays maturation of plants, impairs quality of crop like barley, potato, tobacco, sugarcane and fruits; increases susceptibility to diseases and causes 'lodging' of cereal crops by inducing an undue length thinning of stem inter-nodes.

Nitrogen addition and regulation The control it exerts on soil humus, its complicated transformation, its ready loss by volatilisation and in drainage water, and its notable influence on crops makes it an element of outstanding understanding and attention.

Nitrogen activated by legumes in soil Under most of the circumstances the legumes render the organic matter of the soil more active by adding to it plant residues with a narrow nitrogen-carbon ratio. Such residues are readily decomposed and ammonification soon takes place.

Nitrogen and maintenance of soil humus (1) Since the nitrogen during the rapid dissolution of the soil lignin, and clay, and the carbon is reduced in a more or less definite with it, the amount of soil nitrogen thus determines the amount of organic carbon present. (2) As a definite ratio exists (1 : 1.724) between the organic carbon and the soil organic matter that can be maintained in any soil is, like the carbon contingent upon the amount of nitrogen present. This microbial relationship of the soil nitrogen to the maintenance of these soil organic matters are very powerful and consistent.

Nitrogen conservation in soil When the activity of the heterotrophic bacteria and of the fungi has somewhat abated, and the N : C ratio has become narrower, the microbial demand for nitrogen now becomes available for nitrification, and nitrate accumulation can take place.

Nitrogen control in soils Nitrogen deficits are often met, partially or wholly in four ways-crop residues, farm manure, legumes and commercial fertilizers.

Nitrogen dioxide (NO_2) A brown gas produced by the dissociation of dinitrogen tetroxide, the dissociation being complete at 140°C. Further heating causes dissociation to colourless nitrogen monoxide and oxygen:

$$2NO_2(g) = 2\,NO(g) + O_2\,(g)$$

Nitrogen dioxide can also be made by and action of heat on metal nitrates (not the nitrates of the alkali metals or some of the alkaline-earth metals).

Nitrogen fixation The formation of nitrogenous compounds from atmospheric nitrogen. In nature this may be achieved by electric discharge in the atmosphere or by the activities of certain micro-organisms. The symbiotic bacterial, Rhizobium species are associated with leguminous plants, forming the characteristic nodules on their roots.

Nitrogen fixing bacteria Nitrosomonas and Nitrococcus bacterial.

Nitrogen monoxide (nitric oxide, NO) A colourless gas that is insoluble in water but dissolves in a solution containing iron (II)

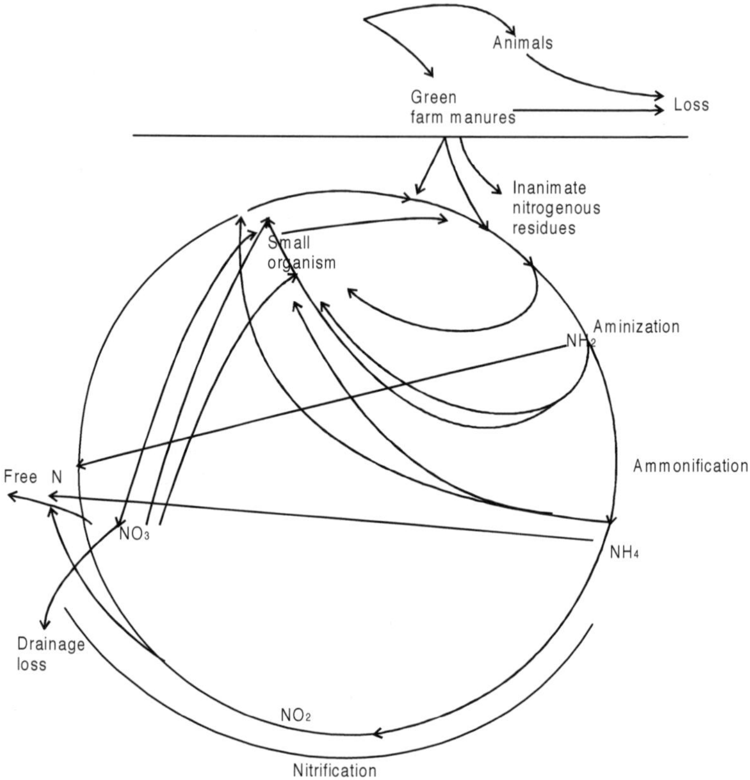

The Nitrogen Cycle of Soils

ions owing to the formation of the complex ion $(FeNO)^{2+}$: the nitrogen monoxide can be released by heating.

$$2NO(g) + O_2(g)–2NO_2(g)$$

Nitrogen and organic matter product in soils It is feasible to provide most of a crop's nitrogen needs in the form of chemically produced fertilizer inputs. In other areas, much of the total nitrogen and organic matter is product of many ecosystems' variables and climatic factors.

Nitrogen ratio with carbon The ratio of nitrogen to carbon ranges in the soil from 1 to 20 or 30 in legumes and farm manure to a high as 1 to 70 or even more in certain strawy conditions. The nitrogen ratio of the bodies of the micro-organisms is not only more constant but much narrower ordinarily falling between 1 : 4 and 1 : 9.

Nitroglycerine A highly explosive substance used in dynamite. It is obtained by treating glycol (1,2,3-trihydroxypropane) with a mixture of concentrated nitric and sulphuric acids. It is not a nitre compound but a nitrate ester.

$$CH_2(NO_3)CH(NO_3)CH_2(NO_3)$$

Nitrophenols ($C_6H_4(OH)NO_2$) Organic compounds formed directly or indirectly by the nitration of phenol. Three isomeric forms are possible. The 2 and 4 isomers are produced by the direct nitration of phenol and can be separated by steam distillation, the 2 isomer being steam volatile. The 3 isomer is produced from nitrobenzene, conversion to 3 nitrophenylamine, and thence by diazotization to 3 nitrophenol.

Nitrous acid (HNO_2) A weak acid known only in solution obtained by acidifying a solution of a nitrite. It readily decomposes on warming or shaking to nitrogen monoxide and nitric acid.

Nitryl ion (nitronium ion) The positive ion NO_2^+.

Nivation (snow patch erosion) The weathering and erosional processes resulting from a snow cover; usually in a hollow on the lee side of a mountain or a ridge, Freeze-thaw is the mot important process.

Noble gases (rare gases, inert gases group of elements) A group of non-atomic, low-boiling gases classified as belonging to group of the periodic table and occupying a position between the highly electronegative group VII elements and the highly electropositive group 1 elements. They are: helium (HeO, neon (Ne), argon (Ar), Krypton (Kr), xenon (Xe) and radon (Rn).

Non-metal Non metal lie in these top right-hand region of the periodic table. They are electronegative elements with a tendency to form covalent compounds or negative ions. They have acidic oxides and hydroxides. In the solid state, non-metals are either covalent volatile crystals or macromolecular (giant-covalent) crystals. *See* also metal, metaloid.

Non-metallic compounds Spergon, Thiosan and Arasan.

Non-polar compound A compound that has molecules with no permanent dipole moment. Examples of non-polar compounds are hydrogen, tetrachloromethane and carbon dioxide.

Non-polar solvent A solvent that does not posses a permanent electric dipole moment and consequently has no drive towards intermolecular association with polar species. Non-polar solvents are usually good solvents for non-polar solutes. For example, carbon tetrachloride, hexane and benzene are good solvents for iodine.

Normal or geologic erosion This is a normal feature of any landscape. Geologic erosion takes place steadily but so slowly that ages are required for it to make any marked alteration in the major features of the earth's surface. There is always an equilibrium between removal and formation of soil so that unless the equilibrium is disturbed by some outside agency the mature soil preserves a more or less constant depth and character indefinitely.

Normality The number of gram equivalents per cubic decimetre of a given solution.

Normal solution A solution that contains one gram equivalent weight per litre of solution. Values are designated by the symbol N, *e.g.* O, 2N, N/10, etc. Because there is not a clear definition of equivalent weight suitable for all reactions, a solution may have one value of normality for one reaction and another value in a different reaction. Because of this, many workers prefer the molar solution notation.

Nucleon number (mass number) Symbol: A, The number of nucleons (protons plus neutrons) in an atomic nucleus.

Nutrients in soil Normally soil is a very big store of food material for human beings. But due to some natural reasons, the production of these materials is gradually decreasing. The main reason is the decreasing of the nutrients of the soil. This depends on the various factors such as kind of the crop grown and climatic conditions of the place.

O

Octadecenoic acid is-9-octadecenoic acid (or oleic acid) is a naturally occurring carboxylic acid present (as glycerides) in fats and oils:

$$CH_3\,(CH_2)_7\,CH:CH(CH_2)_7COOH$$

Octahydrate A crystalline compound containing eight molecules of water of crystallization per molecule of compound.

Octane (C_3H_{18}) A liquid alkane obtained from the light fraction of crude oil. Octane and its isomers are the principle constituents of petrol, which is obtained as the refined light fraction from crude oil. *See* also octane rating.

Octavalent Having a valency of eight.

Ohm The SI unit of electrical resistance, equal to a resistance that passes a current of one ampere when there is an electric potential difference of one volt across it. 1 – 1 VA^{-1}. Formerly, it was defined in terms of the resistance of a column of mercury under specified conditions.

Oil Any of various viscous liquids. *See* petroleum, glyceride.

Oil refinery The industrial plant at which crude oil is separated into fractions in a vertical container to produce diesel oil paraffin petrol and refinery gas. The residue is used to produce lubricating oils, paraffin wax, bitumen and a range of chemicals.

Oil sand (tar sand) An open-textured sand or sandstone in which oil has accumulated in the pores in the rock, thus forming an oil reservoir.

Oil shale A fine-grained sedimentary rock, normally finely layered, containing organic matter from which oil can be extracted by heating. Free flowing oil is unusual although the rock may contain pockets or small veins of oily substance. Oil yields of 38 litres per metric ton to 550 litres per metric ton have been recorded.

Oil of vitriol Sulphuric (VI) acid.

Old and new groups of alluvial soils The alluvial group has been divided into two broad geological sub-divisions, *i.e.*, old land new alluvium. The old alluvium, locally called 'bangar', is in the process of denudation, but the new alluvium, known as 'khader', is in the process of building up.

Oleate A salt of ester of oleic acid.

Olefin *See* alkane.

Oleic acid An unsaturated fatty acid occurring as the glyceride in oils and fats. Oleic acid occurs naturally in larger quantities than any other fatty acid. In many organisms oleic acid can be synthesized directly from stearic acid and further enzymatic paths exist for conversion to linoleic acid and linolenic acid. This pathway does not occur in man and the higher animals so plant sources are an essential dietary element.

Olivine Olivine is a ferro-magnesium silicate, and serpentine is hydrated magnesium silicate. It is an iron carrying mineral. On the assumption of water serpentine, a hydrated product, and ferrous may result. The latter is an oxidized to ferric state.

$$3MgSiO_4 + 2H_2O = H + Mg_3Si_2O_9 + SiO_2 + 3Fe$$

$$4FeO + 3H_2O + O_2 = 2Fe_2O_3\ 3H_2O$$

Onion weathering *See* exfoliation.

Optical isomerism *See* isomerism, optical activity.

Optimum Most favourable condition.

Optimum concentration of nutrients in soils It is known that a deficiency of nitrates, phosphates, potassium and iron restricts growth and that a soil is seldom so poor in these nutrients as to be totally in capable of supporting plant life. It is also a common knowledge that an excessive concentration of such substances as sodium chloride, sulphate, carbonate and bicarbonate inhibits crop growth almost completely.

Ore The deposit from which a mineral is obtained. An ore contains sufficient useful mineral material to make it profitable to extract it. In some ores, several metals are associated together, *e.g.* silver and mercury are associated with lead and zinc ores. The existence of a second metal in an ore may make its working profitable.

Organic chemistry The chemistry of compounds of carbon. Originally the term organic chemical referred to chemical compounds present in living matter, but now it is any carbon compound with the exception of certain simple ones, such as the carbon oxides, carbonates, cyanides and cyanates. These are generally studied in inorganic chemistry.

Organic deposits There are three kinds of these deposits–(1) sedimentary peats which usually accumulates in comparatively deep water and therefore generally found well down in the profile. These soils are unsatisfactory of agriculture, (2) the group of fibrous peat occur often in the same swamp deposit. These are heterogeneous in character, high in water holding capacity and may exhibit varying degrees of decomposition, and (3) since trees are the climax vegetation in swamp deposits, woody peat, when it occurs, is usually at the surface of the organic accumulation.

Organic matter with large amount of carbon When such a matter and condition is present as improportion to the nitrogen present in the soil, the nitrification and ammonification processes are not only retarded but may readily be reversed temporarily. The soil microorganism having large amounts of energy-producing materials at this deposit, multiply rapidly and use the nitrogen themselves, thus interfering with its simplification and appearance as ammonium and nitrates.

Organic materials Besides inorganic substances, soils contain organic matter in amounts ranging from less that 1 per cent in arid sandy soils to as much as 90 per cent in peaty soils. The presence of organic matter distinguished soil from disintegrate rock material.

Organic mercury compounds Ethyl mercury phosphate, ethyl mercury chloride and chlorophenol mercury.

Organic material absorption The soil organic matter may, to some extent, serve as a direct source of plant nutrients, providing not only nitrogen and sulphur, but also carbon, oxygen and hydrogen elements usually considered as coming from air and water. Since these ingredients are absorbed in complex molecular condition, a great saving of energy is effected, as they become a part of cell contents with less synthetic activity than would be the case with simpler nutrients.

Organic soils These have lighter dry weight, colour is brown or dark-brown, water holding capacity is high. On volume basis, a cubic foot of peat soil at optimum moisture can supply somewhat more water to plants than a representative material soil also at optimum. An organic soil has invariably a better physical condition. On the basis of their organic content, soils of two kinds are recognised–(1) mineral, and (2) organic. The mineral soils vary in organic matter from a mere trace to as high as 12 to 15 per cent. Organic soils, possesses much more organic

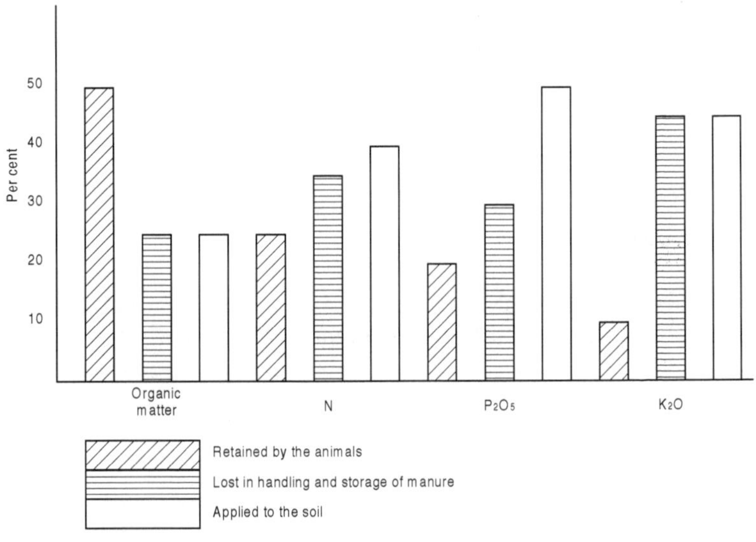

An approximate estimate of proportion of the organic matter, nitrogen, phosphoric acid and potash present in the food

material, the content of which ranges from 20 to as high as, in some cases 90, 95 per cent.

Organic weathering The decomposition and disintegration of rocks by plants and animals. This involves both mechanical and physical weathering. For example, mosses and lichens retain rainwater and humic acids in contact with rock surfaces; this enables chemical weathering to break down the rock under the plants. The result of the process can be seen in the shallow depressions that form beneath mosses and lichens. Burrowing animals and the penetration of plant roots can cause the physical weathering of rocks.

Origin of alkali in soil In addition of carbonate of soda, black or brown alkali may also be formed by calcium chloride or sodium nitrate. A well developed solonetz is very strongly alkaline in reaction and its B horizon is characterised by a column of prismatic structure.

Other soil climatic areas These are represented by the desert soils, grey and brown soils prairies and chernozem soils of the sub-montane Tarai, and laterites of the hot and humid.

Over-liming The drastic rise in pH after adding overdose of lime to the soil might for a time immobilize iron or manganese thus creating a temporary deficiency of one or both of these nutrients. Copper and zinc may also be affected in the same way. With boron the problem becomes more complicated.

Oxalic acid (ethanedioic acid) A dicarboxylic acid, which occurs in rhubarb leaves, wood sorrel and the garden oxalis (hence the name), oxaloacetic acid (OAA), a water-soluble carboxylic acid, structurally related to fumeric acid and maleic acid. Oxaloacetic acid forms part of the citric acid cycle, it is produced from L-malate in an NAD requiring reaction and itself is a step towards the formation of citric acid in a reaction involving pyruvate ion and coenzyme A.

Oxidant An oxidizing agent. In rocket fuels, the oxidant is the compound that provides the oxygen for combustion (*e.g.* liquid oxygen or hydrogen peroxide).

Oxidase An enzyme which promotes oxidation.

Oxidation An atom, an ion or a molecule is said to undergo oxidation or to be oxidized when it loses electrons. The process may be affected chemically, *i.e.* in which case oxidation occurs at the anode. For example,

$$2Na + Cl_2 - 2Na^+ + 2Cl^-$$

Where chlorine is the oxidizing agent and sodium is oxidized,

and $$4CN^- 2Cu^{2+} - C_2N_2 + 2CuCN$$

where, Cu^{2+} is the oxidizing agent and CN^- is oxidized. The oxidation state of an atom is indicated by the number of electrons lost or effectively lost by the neutral atom, *i.e.* the oxidation number of a negative ion is negative. The process of oxidation is the converse of reduction. *See* also redox.

Of the various chemical changes due to weathering, oxidation is usually one of the first to be noticed. It is particularly manifest in the rocks carrying iron in the sulphide. Carbonate and silicate forms oxygen combining with various minerals produces soluble oxides, which on dissolving in water weaken the rocks and bring about their disintegration.

$$2\,HNO_3 + M - + 2NO_2 + H_2O; MO + 2HNO_3 - H_2 + M(NO_3)_2$$

Oxidation and soil formation Oxygen combining with various minerals produces soluble oxides, which on dissolving in water weaken the rocks and bring about their disintegration.

Oxide of lime It is commonly known as the bumed lime or quicklime or often as simple oxide. It is sold in two forms–(1) ground burned, and (2) lump lime. It is quite caustic.

Oxime A type of organic compound containing the C : NOH group, formed by reaction of an aldehyde or ketone with hydroxylamine (NH_2OH).

Oxygen An element essential to living organisms both as a constituent of carbohydrates, fats, proteins and their derivatives, and in aerobic respiration. It enters plants both as carbon dioxide and water, the oxygen from water being released in gaseous form as by-product of oxygen. This is an essential element as it helps in building tissue in plants. In nature, plants meet their full requirement from the environment they grow in.

Oxygen debt A physiological state that occurs when a normally aerobic animal is forced to respire anaerobically during a temporary shortage of oxygen (anoxia), *e.g.* due to violent muscular exertion.

Oxygen quotient (QO_2) The rate of oxygen consumption of an organism or tissue. It is usually expressed in microliters of oxygen per milligram of dry weight per house. Small organisms tend to have higher oxygen quotients than larger ones.

Ozone (trioxygen, O_3) An allotrope of oxygen made by passing oxygen through a silent electric discharge. Ozone is unstable and decomposes to oxygen on warming. It is present in the upper layers of the atmosphere, where it screens the earth from harmful short-wave ultraviolet radiation.

It is strong oxidizing agent. Its concentration can be determined by reacting it with iodide ions and titrating the iodine formed with standard sodium thiosulphate (VI). It forms ozonides with alkenes. *See* ozonolysis.

Ozone layer (ozonosphere) A zone within the atmosphere, generally between 20 and 50 km, where ozone forms in its greatest concentrations. It results from the splitting of oxygen (O_2) molecules by the ultraviolet radiation from the Sun to form atomic oxygen (O), which then combines with other oxygen (O_2) molecules to give ozone (O_3).

P

Palaeontology A study of fossils including their form and evolution, their place in space and time and their associations with other fossils. It includes the study of both the fossilized remains of the organisms themselves and trace fossils such as the tracks made by organisms.

Paraffin immersion method of determining volume weight This method is useful for heavy soils. Small pieces of soil are dried, weighted and then coated very thinly with paraffin, just sufficiently to prevent the entrance of water, yet not enough to enter serious experimental error. The weight of water displaced by a number of such pieces may be determined easily by the use of a graduated cylinder.

Paraffin wax A solid mixture of hydrocarbons, obtained from petroleum.

Paraformaldehyde *See* methanol.

Parallelogram law of forces If a body is acted upon by two forces simultaneously, represented by the two adjacent sides of a parallelogram in magnitude and direction, then their resultant can be represented in magnitude and direction by the diagonal of the parallelogram passing through that point.

Parallelogram law of velocities If a body possesses two uniform velocities simultaneously, represented by the two adjacent sides of a parallelogram in magnitude as well as direction, then their resultant can be represented in magnitude and direction by the diagonal of the parallelogram passing through that point.

Parent material The weathered and partially broken rock from which a soil is formed. This may not be the same as the underlying bedrock, for example, it may be a deposit of glacial till or alluvium. Different rock types produce different soils; for example, those formed over limestones are base rich and stone free. In many soil profiles, parent material dominates the C horizon, whereas the B horizon has a mixture of mineral and organic (humus) fractions. Parent material and climate are two most influential factors in soil formation. Parent materials are particularly important in areas of varied surface geology. They govern the soil's chemical quality through the precise composition of minerals, and also affect physical properties such as soil texture.

Peadalfer A mature free-draining zonal soil of humid climates in which leaching has removed base compounds (*e.g.* calcium) so that the soil becomes generally acidic. Pedalfers are rich in aluminium and iron. Examples include podzols, brown earths, latesoils and prairie soils. Compare pedocal. *See* also acid soil.

Peat The dark soft fibrous or structureless partially decomposed remains of plants forming accumulations over 50 cm in depth. Peat is formed when vegetation dies in waterlogged anaerobic (airless) conditions and the decomposing activity of small organisms is hindered by the lack of free oxygen.

Peat soils colloidal nature Because of the very high organic matter and the particular type of humiciation, the colloidal nature of peat soils is striking far beyond that exhibited by mineral soils colloid.

Peat soil composition Peat soils are more variable than the mineral soils. Certain outstanding and fairly constant characteristics in evidence can be shown as thus:

Constituents	Peat Surface Soil (Expressed as %)	Mineral Surface Soil (Expressed as %)
Organic matter	80.0	4.0
Nitrogen (NO_2)	2.50	0.15
Phosphoric acid (P_2O_5)	0.20	0.110
Potash (K_2O)	0.10	2.0
Lime (CaO)	4.00	0.60
Manganese (MgO)	0.50	0.50
Sulphur trioxide (SO_3)	1.50	0.10

Peat soils of lakes These are divided in two groups–(1) Undecomposed or only slightly decomposed soils, and (2) Those are moderately or definitely decayed. These soils are important source of organic manure. The leaves and other plant material and animal residues that get into the lake transform themselves into an important manure. Peat soils about 0.95 to 1.04 per cent in nitrogen is present. It contains 8 per cent nitrogen and 3 per cent phosphorus.

Peat soils' physical characteristics The colour of typical cultivated peat soil, dark brown or intensely black when it is wet, is perhaps the first physical characteristic that attract the attention. It has a light weight also. The volume weight compared with mineral surface soils is surprisingly low, 0.20–0.30 perhaps being a fair estimate. It has high water holding capacity. A typical woody or fibrous peat soil is always having an ideal physical condition.

Percolation the downward movement of water through the pores of a rock or soil.

Peat soils' profile sequence One might expect that the profile succession of peat soil would be rather definite for any region. Actually, no regular sequence is found, due to large part to the irregular and uncoordinated fluctuations in the water levels of swamps and bogs.

Peat soils' total exchange capacity Since the buffering of a soil is determined in large degree by the magnitude of its total exchange capacity, peat soils in general show an unusually marked resistance to a change in pH, much greater than do mineral soils.

Pebble A rounded fragment of rock; it is larger than a gravel stone but smaller than a cobble. It is sometimes defined as having a diameter of 4-6.4 mm. The rounding of the rock fragment may result from either wind or water action. Large numbers of pebbles that accumulate in one place form a pebble bed.

Pedology The scientific study of the characteristics, development, and distribution of soils.

Pedon A column of soil extending from the ground surface down to parent material, used in field investigations of soils. As it is three-dimensional a pedon can complement the conventional two-dimensional soil profile by showing. horizontal changes through a soil horizon. Compare soil profile.

Percentage base saturation in soils This is a proportion of bases (*i.e.* Ca, Ma, K etc.) to hydrogen in the exchange complex. This relationship ordinarily is expressed in terms of percentage base saturation.

Percentage slope and length of slope The speed and extent of run-off depends of the slope of the land. The greater the slope, the greater the velocity of flow of the run-off water. According to the law of falling bodies, velocity varies as the square root of the vertical drop. Hence, if the land slope is increased four times, the velocity of the water flowing on the slope is approximately doubled.

Periodic acid *See* iodic (VII) acid.

Periodic law The law upon which the modern periodic table is based, enunciated in 1869 by Mendeleev. This law stated that the properties of the elements are a periodic function of their atomic weights, *i.e.* if arranged in order of increasing atomic weight then elements having similar properties occur at fixed intervals. Certain exceptions or gaps in the table lead to the view that the nuclear charge is a more characteristic function, thus the modern statement of the periodic law is that the physical and chemical properties of elements are a periodic function of their proton number.

Periodic table A table of the elements arranged in order of increasing proton number to show similarities in chemical behaviour between elements. Horizontal rows of elements are called periods. Across a period there is a general trend from metallic to non-metallic behaviour. Vertical columns of related elements are called groups.

Permanent wilting point The point at which soil has dried to the extent that plants can no longer remove the remaining water held on the soil particles, and begin to wilt.

Permeable rock A rock through which water can pass more or less freely from the upper surface to the lower surface. It may be either porous (*e.g.* sandstone, colitic limestone) or pervious (*e.g.* chalk). Water contained within a permeable rock is capable of being extracted by pumping. *See* also pervious rock, porosity.

Permutit (Trade name) A substance used to remove unwanted chemicals that have dissolved in water. It is zeolite consisting of a complex chemical compounds, sodium aluminium silicate (Na^+ Al silicate$^-$). When hard water is passed over this material, calcium and magnesium ions exchange with sodium ions in the permutit. This is a good example of ion exchange.

Peroxide 1. An oxide containing the O^--O– ion.

2. A compound containing the -O-O- group.

Pervious rock A rock through which water can pass freely as a result of joints, bedding planes, cracks, and fissures in the rock. For example, some lime stones (*e.g.* chalk) and igneous rocks. Petrochemicals are obtained as by-products of crude oil or natural gas. A large number of goods used today are derived from petrochemicals, *e.g.* fertilizers, plastics, synthetic fibres such as nylon, pharmaceuticals, and synthetic rubber.

Petrology The study of all aspects of rocks, including mineral composition, texture, structure, origin, occurrence, alteration and relationship with other rocks.

pF Energy values are not expressed in centimetre heights of unit water columns, but as a logarithm of such figures. This is called for convenience pF.

pF and capillary water The rate of adjustment under given tensions is influenced by the size and continuity of the pore spaces and by the moisture content of the soil mass invaded by the capillary water front. Large capillary pore spaces, such as are likely to be present in sandy soils, encourage a comparatively rapid adjustment but the presence of macro-interstices provide discontinuities which may greatly impede such movement.

pF and hygroscopic coefficient If hygroscopic coefficient represent a maximum adsorption, and atmosphere nearly saturated should be employed. Many experiments favour the use of a 3.3 per cent solution of sulphuric acid which gives a relative humidity at laboratory temperature of 98.2 per cent and a pF of 4.5.

pF and moisture equivalent Due to the conditions the short column of soil is subjected to as the pressure applied, the average tension is equivalent to a pull on the free column of only 55 times the force of gravity. Hence the moisture equivalent corresponds to a pF of 2.7 or about 1/2 atmosphere.

pF value and field capacity The retentive capacity of a column of soil for water may be determined by fully saturating it and then allowing to drain undisturbed by evaporation. The amount of water thus retained, known as field capacity, will depend upon a number of conditions. The length of the column of soil is an important factor, the greatest amount of moisture being retained at the base of the column.

pF and vapour pressure in soil Different degrees of air saturation of vapour pressure develop different pF values for the moisture absorbed. At 50 per cent air saturation the corresponding pF is about 6; at 75 per cent it is about 5.6; at 90 per cent about 5 and at 98 per cent is 4.5.

pF graphs By pF studies it has been noted that the pF curves for the maximum hygroscopic coefficient for mineral soils falls at or near a pF of 4.5, equivalent to a pressure of about 31 atoms.

pF limits The soil solution is coexistent with the part of the liquid water that remains in the soil after free drainage has taken place. Hence it exhibits an energy range from pF 4.5 to pF 2.7 in case of capillary water.

pF limits of capillary water movement Roughly speaking pF 1.6 represents the transition from capillary to non-capillary pores the moisture below pF 1.6 moving largely by gravity. Between pF 2.7 and 1.6 water may be affected definitely by both capillary and gravity, but above pF 2.7 the adjustment is almost entirely of a capillary nature.

pF of soil Some of the soil water is retained with such a tremendous tenacity that the suction of pulls necessary to remove it, if expressed directly in centimetre heights of water gives awkwardly larger figures. Certain increments of the colloidally absorbed moisture require for their removal a force more than equivalent to the weight of a column of water 1,000,000 centimetres in height. This exceeds by several thousand feet the height of our loftiest mountains. In order to avoid such inconvenient figures and to make the plotting of curves possible, energy values may be expressed, not in centimetre heights of unit water columns but the pF.

pH A measure of the acidity or alkalinity in a soil or solution. The scale ranges from 0 (completely acid) to 14 (highly alkaline), the median value pH-7 (neutral) being that of pure water. The extremes of acidity (below pH-3) and alkalinity (above pH-11) seldom occur naturally. The optimum for many crops is pH-6 (*i.e.* slightly acid). The pH is given by 10_{10} $(1/H^+)$, where (H^+) is the hydrogen ion concentration in moles per litre. It should be stressed that the pH value of a soil indicates only the strength or intensity of acidity, just as the degrees of temperature indicate the intensity of heat. The pH value tells nothing about the total acidity of soil remedial measures required for correcting soil acidity in any particular soil.

pH and buffering It is important for two cogent reasons–(1) the stabilization of soil pH, and (2) the amounts of amendments necessary to affect a given alteration in soil reaction.

pH and colloidal control Before any satisfactory control measure be made in respect to the soil acidity, or for that matter soil reaction in general. The control mechanism of soil pH is colloidal complex, which according to its nature and condition, exerts a dominating equilibrium restraint and regulation upon the reaction of the soil solution.

pH and disease prevention of plant Prevention is the best defence from diseases produced by soil organisms. Regulation of pH is effective to a certain extent with a potato scab and the club-rot of cabbage.

pH and lateritic soils Typical lateritic soils are low in total exchange capacity compared to temperate regions' soils.

pH correlation with active forms of nutrients When the exchangeable calcium and magnesium are lost by leaching, the acidity of the soil gradually increases. This shows that in anyone soil, there is definite correlation between the pH and the activity of these constituents.

pH correlation with phosphate ions When the soil is distinctly alkaline, the PO_4 ion apparently is the commonest form present. This ion is acquired rather slowly by plants from most soils. But as the pH is lowered and becomes slightly acid, the PO_4 ions prevail. At higher acidity H_2PO_4 ions tend to dominate.

pH determination and measurement The most accurate method of measurement of pH is the method of potentiometer, and electrometric method in which the H-ion concentration of the soil solution is balanced against a standard hydrogen electrode or an electrode that functions in a similar way.

pH determination with dye method This method works satisfactorily when the soil is air dry, as a better contact is established between the soil and the indicator.

pH determination with indicators This method is very simple but less accurate than electrometric and consists in the use of certain indicators.

pH major changes in soils The most important and persistent cause of change in the reaction of the solutions of soils in general is the carbon dioxide. A general effect of this kind can be shown here.

pH minor changes The drying of soil, especially above field temperatures, will often cause a noticeable increase in acidity. This may be due to some change in the organization of the colloidal matter and should be kept in mind in preparing oil samples for pH.

pH nutritional importance H-ion have a considerable importance for–(1) the solubility but also upon, and (2) the facility with which nutrients, even when readily soluble, are absorbed and used by plants. Iron, manganese and zinc are such examples of the relationship as they tend to become less and less soluble as the pH rises towards 8.

pH of aridic soils In aridic soils especially if they have sodium carbonate pH may be about 8 or more.

pH of soils It is important for the reaction of soil solutions *i.e.* whether the soil solution is acidic or alkaline, or neutral. Some show the reserve of this (OH) and are alkaline. The exact relationship in any particular case is usually evaluated in terms of H-ion concentration and expressed logarithmically in pH values.

pH range of soils It is often required to interpret the pH of the soil solution in general and comparative terms:

Very alkaline above	8.0	Moderately acid	5.5–5.9
Alkaline	7.4–8.0	Strongly acid	5.5–5.4
Neutral or so	6.6–7.3	Very strongly acid	4.3–4.9
Slightly acid	6.0–6.5	Extremely acid below	4.3

pH reaction in relation bacteria Moulds thrive in soils of high acidity and low lime. The response of most bacteria is quite the reverse although there are some exceptions. At a pH below 5.5, the average mineral soil is too low in active calcium to support vigorous bacterial activity.

pH value of a solution This is the value in solution of the logarithm of the reciprocal of the H-ion concentration, it may be expressed

as a: $$pH = \log \frac{1}{[H^+]}$$

the small p indicates that the values are logarithmic while the capital H is a remainder that the hydrogen concentration is being measured. As the H-ion concentration increases, the pH value decreases and vice versa. The change in H-ion concentration is geometric and not arithmetic since there are 9 acid units between pH 6 and 7,90 between pH 5 and 6, and 900 between pH 4 and 5 and so on. Alkalinity is expressed in terms of H-ion concentration, a pH greater than 7 indicates an alkaline reaction.

Phase equilibrium A state in which the proportions of the various phases in a chemical system are fixed. When two or more phases are present at fixed temperature and pressure, a dynamic condition will be established in which individual particles will leave one phase and enter another; at equilibrium this will be balanced by an equal number or particles making the reverse change, leaving the total composition unchanged.

Phenol 1. (carbolic acid, C_6H_5OH) A white crystalline solid used to make a variety of other organic compounds. 2. A type of organic compound in which a hydroxyl group is bound directly to one of the carbon atoms of an aromatic ring. Phenols do not show the behaviour typical of alcohols. In particular they are more acidic because of the electron-withdrawing effect of the aromatic ring. The preparation of phenol itself is by fusing the sodium salt of the sulphonic acid with sodium hydroxide:

$$C_6H_5SO_{2-}\ ONa + 2NaOH - C_6H_5ONa + Na_2SO_3 + H_2O$$

Phenolphthalein An acid-alkali indicator that is colourless in acids and red in alkali. It has a pH range from 8.4–10.0 and is a frequently used indicator for the detection of pH change in acid-alkali titrations. Phenolphthalein, together with borax, is used to test for saccharide derivatives turn the solution from red to colourless but on boiling the red colour returns.

Phenylhydrazine A colourless liquid that reacts with aldehydes and ketones to give phenylhydrazine. These are white solids with definite melting points that can be used to identify the respective aldehydes and ketones. With monosaccharides, phenylhydrazine forms ozazones, yellow crystalline compounds, that are distinctive for most monosaccharides. *See* ozazones.

Phenyl groups The group C_6H_5, derived from benzene.

Phosphagen Creatine phosphate.

Phosphatide A phospholipid.

Phosphate absorption The nature of the phosphate ions varies with pH of the soil solution. When the soil is distinctly alkaline, the PO_4 ion is the common form. This ion is acquired, rather slowly by plants in most soils. But as pH is lowered, and the soil becomes slightly to moderately acidic, the HPO_4 and H_2PO_4 ions prevail. At higher acidities H_2PO_4 tend to dominate. These two forms are thought to be absorbed more easily by higher plants and probably by micro-organisms also.

Phosphate fertilizers These are classified as natural phosphates, treated or processed phosphates, and by-product phosphates and chemical phosphates. The ordinary grades (16 to 20 per cent of available P_2O_5) are made by treating rock phosphate with suitable amounts of sulphuric acid. A large proportion is thus changed to the monocalcic condition [$Ca_2H_4(PO_4)_2$] although some is left in the dicalcic condition [$Ca_2H_2(PO_4)_2$]. Much of the superphosphate now used is this ordinary grade, which consists of about 31 per cent phosphates, 50 per cent gypsum and 19 per cent impurities of various kinds. The total phosphorus in the bulk is low, ranging from 7 to 9 lbs per hundred weight, while sulphur and calcium amount to about 12 to 18 lbs respectively. There are two grades of high-grade superphosphates, the double (32 per cent available P_2O_5) and the treble (40–48 per cent available P_2O_5).

Phosphate management of soils *See* soils phosphoric acid availability.

Phosphatic manures Such manures which have phosphorus as the main constituent in them such as superphosphate (single, double and triple) etc.

Phosphorescent glow Same as phosphorescence.

Phot A unit of illumination in the c.g.s. system, equal to an illumination one lumen per square centimetre. It is equal to 10^4 lx.

Photo Light Phreatic water Ground water, especially that below the level of the water table. Compare vadose water.

Physical range A change to a substance that does not alter its chemical properties. Physical changes (*e.g.* melting, boiling and dissolving) are comparatively easy to reverse.

Physical chemistry The branch of chemistry concerned with physical properties of compounds and how these depend on the chemical bonding.

Physical properties of soils The physical properties of soils, such as water-holding capacity, permeability to water pore space, aeration, plasticity, cohesion etc., are determined by the size and arrangement of particles composing them. The relative proportions in which particles of different size groups occur in a soil are ascertained by mechanical analysis. International method is a lengthy process. To shorten it, use is now made in some countries of a special kind of hydrometer. By placing this instrument in a sedimentation cylinder containing the soil in suspension and taking readings at particular time intervals, the percentage of particles of various size groups can be easily determined.

Physical status of black soils They are extremely soft when wet, but on drying they contract markedly, break up into hard baked blocks and develop fissures, which may be 2 to 4 inches wide and 1 to 4 feet deep. Medium deep soils overlying disintegrating murum are well drained. However, all black soils are highly retentive of moisture.

Physical weathering *See* mechanical weathering.

Physiographic sub-divisions of soils Depending on the mode of genesis, several physiographic sub-divisions are recognisable within a group such as (1) sandy gravely, (2) residual, (3) partly residual and partly alluvial, (4) loaves (5) clay-loaves etc.

Pitch (of a scale) Pith of a screw gauge is the perpendicular distance between its two consecutive threads. When the screw is given one complete rotation, the distance moved forward or backward by the screw on the linear scale is equal to its pitch.

Pitchblende A form of uraninite, an ore of uranium. Pitchblende is the most important source of uranium. It is found chiefly in hydrothermal veins associated with small amounts of other elements such as lead, thorium and cerium.

Plants in soil formation Simple plants such as mosses and lichens grow upon exposed rock, there to catch dust until a thin layer of highly inorganic material is accumulated. Higher plants sometime exert a prying effect in rock which results in a disintegration.

Planosol An intraxzonal soil of poorly drained areas, *e.g.* flat areas. Planosols are characterized by the eluviation of fine materials to form a claypan.

Plant succession *See* plant science.

Plaster of Paris *See* calcium sulphate.

Plasticity From the practical standpoint, certain other characteristic of clay assume some considerable importance. There are three physical conditions: plasticity, cohesion and flocculation. More water and less soil produces a sol and less and much more soil produces gel.

Plasticity of soil Plasticity is the property that enables moist soils to change shape on the application of force and retain this shape even when the force is removed. Soils with very low plasticity, such as sands or loamy sand, form a cast when squeezed in moist condition but the cast crumbles on touching.

Platinum A transition metal. It occurs naturally in Australia and Canada, either free or in association with other platinum metals. It is resistant to oxidation of ammonia to manufacture nitric acid and in jewellery. Symbol: pt; m.p. 1770°C; b.p. 3830°C; r.d. 21.5; p.n. 78; r.a.m. 195.09.

Platinum black A finely divided black form of platinum produced, as a coating, by evaporating platinum onto a surface in an inert atmosphere. Platinum-black coatings are used as pure absorbent electrode coatings in experiments on electric cells. They are also used, like carbon black coatings, to improve the ability of a surface to absorb radiation.

Platinum metals The group of transition metals ruthenium (Ru), osmium (Os), rhodium (Rh), iridium (Ir), palladium (Pd), and platinum (Pt). They are sometimes classed together as they have related properties and compounds.

Platy In this structural type the aggregates are arranged in relatively thin horizontal plates, leaflets or lenses. Such an aggregation usually is confined to the A horizon of humid region soils.

Pleistocene age This is the ice age when weather not only brought a lowering or temperature and an increase in precipitation but established such a relation between them that the heat of summer could not melt the winter accumulation of snow.

Pleroylglutamic acid *See* folic acid.

Pliocene The last geological epoch of the tertiary period. It followed the Miocene and extended for about two million years to the beginning of the Quarteary period. During this time the conditions that prevailed throughout the Miocene continued in that the sea continued to withdraw its previous areas and the temperatures continued to drop.

Plumbane (lead (V) hydride, PBH_4) A colourless unstable gas that can be obtained by the action of acids on a mixture of magnesium and lead pellets.

Plutonic rock An igneous rock that cools in a large mass at great depth within the earth's crust. Cooling takes place very gradually at great pressure resulting in coarse-textured rock in which the crystals are large and about the *Sqme* size. Examples include granite, diorite, gabber and peridodotite. Compare hypabyssal rock, extrusive rock.

Podsol The type of soil found under heathland and coniferous forests in temperate climates. It is strongly acid and often deficient in nutrients as a result of leaching.

Podzol humus When organic matter is more abundant, or rainfall heavier, layer of precipitated humus, very fine and smooth in feel, accumulates between the leached zone (A_2) and the B_2. This humus horizon is a true B_1 and if present in any amount, gives what is called humus podzol.

Podzol iron In many podzols the B_1 horizon is very faintly developed, the A_2 apparently resting directly on the B_2, which is especially high in iron. Such a profile is called iron podzol.

Podzolic soils Strongly leached surface soils over clay pans.

Podzolization Accompanied with accumulation of organic matter of vegetation caused by a type of decomposition that will develop organic matter yielding acids other substances of high solvent capacity. When podzolization is intense fungi seem to be the micro-organisms engaged in the production of such humus, bacterial activity being at minimum. In such an acid, low calcium medium, earthworms are practically absent. pH values are as low as 3.5 are often found in this humus, especially the H-Iayer. Not only climatic conditions should be cool and damp, but also sufficient excess moisture must be present, at least at certain times to promote leaching.

Pollution It is the fouling of the environment by man which is harmful for living organisms or reduces all the living facilities provided by the society from time to time. Exhaust fumes from cars, sulphur dioxide from industrial chimneys, chemicals such as fertilizers and the dumping of wastes, and litter and water are responsible for bringing about pollution in human dwellings.

Polonium A radioactive metallic element belonging to group VI of the periodic table. It occurs in very minute quantities in uranium ores. Over 30 radioisotopes are known, nearly all alpha particle emitters. A suggested use of polonium-210 is as a thermoelectric power source in satellites. Symbol: Po; m.p. 254°C; b.p. 962°C; r.d. 9; p.n. 84; stablest isotope^{209} Po.

Polycyclic Describing a compound that has two or more rings in its molecules.

Polyene An alkene with more than two double bonds in its molecules.

Polyethene (polyethylene, polythene) A synthetic polymer made from ethane. It is produced in two forms; a soft material of low density and a harder, higher density form, which is more rigid. It can be made by the Siegler process.

Polyhydric alcohol An alcohol that has several -OH. groups in its molecules.

Polymer A compound in which there are very large molecules made up of repeating molecular units (monomers). Polymers do not usually have a definite relative molecular mass, as there are variations in the lengths of different chains. They may be natural substances (*e.g.* polysaccharides of proteins) or synthetic materials (*e.g.* nylon or polyethene). *See* also polymerization.

Polymerization The process in which one or more compounds react to form a polymer. Homopolymerization is polymerization of lone monomer (*e.g.* the formation of polyethene). Copolymerization involves two or more monomers for nylon. Addition polymerization occurs when the monomers undergo additional reactions with no other substance formed. Condensation polymerization involves the elimination of small molecules in formation of the polymer.

Polypeptide A compound that contains many amino acids linked together by peptide bonds. **Polypropene** (or polypropylene) A synthetic polymer made from propane. It is similar in properties to polythene, but stronger and lighter. The propane is polymerized by the Ziegler process.

Polysaccharide High molecular polymers of the monosaccharide or sugar. They contain many repeated units in their molecular structures. They can be broken down to smaller polysaccharides, disaccharides and monosaccharide by hydrolysis or by the appropriate enzyme. Important polysaccharides are insulin, strarch, glycogen (also known as animal starch), and cellulose. *See* also carbohydrate, sugar.

Pore size and shape Pores of marosoze lie between the larger soil particles and granules and usually are connected by interstices and passages much smaller in diameter. Their arrangement is by no means regular although they often tend to occur in vertical strings, their outline and connections are twisting beyond any definite description.

Pore space amount As might be expected, considerable difference in total pore space occurs depending upon condition. As sandy surface soils show a range from 35 to 50 per cent while heavy soils vary from 40 to 60 per cent or even more depending on the high organic matter and marked granulation. The handling of a soil exerts a marked influence upon the pore space of the soil.

Pore space calculation Per cent pore space =

$$100 = \frac{\text{(Volume weight)}}{\text{Specific gravity}} \times \frac{100}{1}$$

Pore space in soil The space between particle to particle of soils is the pore space of the soil. Pore space of soil is occupied by air and water in constantly varying proportions. The amount of the pore space is determined largely by structural conditions, *i.e.* by the interrelated influences of texture, compaction and aggregation.

Pore space types Two types of pore space occur in the soil-macro and micro. While no sharp line of demarcation occurs, the former characterically allow the ready movement of air and percolation of water. The equilization of gases in the latter type of pore is greatly impeded while movement is restricted largely to slow adjustment. Thus is a sandy soil, in spite of surprisingly rapid due to the dominance of the macro interstices.

Porosity The ratio of the amount of space between the mineral grains to the total volume of a rock soil, or sediment. It is usually expressed as a percentage of the volume. For example, the porosity of an evenly graded rock with perfectly packed grains is 27 per cent. Sediments vary a great deal from less than 1 per cent to over 50 per cent. Sandstones range from 5–15 per cent clays, which are very porous, have a porosity reaching 50 per cent.

Potash A salt of potassium of a mineral containing potassium. Salts such as potassium chloride or potassium magnesium chlorate are found in association with common salt at sites of salt lakes, for example the Dead Sea.

Potash from other sources Wood ash, cattle dung ash, leaf mound, tobacco stems and water hyacinth are the available indigenous sources of potash. Unleached wood ash contains 5 to 6 per cent of potash in the form of potassium carbonate (which is alkaline in reaction), 1 to 2 per cent phosphoric acid and 25 to 30 per cent lime (CaO). Both the potassium carbonate and lime in the wood ashes counteract acidity in the soil. Groundnut shell, paddy husk and bagasse ashes are available near decorticating factories and rice and sugar mills. These also contain a fair amount of potash and some phosphoric acid. Groundnut tobacco stems contain 2 to 3 per cent of nitrogen and 6 to 10 per cent of potash in quickly available forms.

Potassium One of the essential elements in plants and animals. It is absorbed by plant roots as the potassium ion, K^+, and in plants is the most abundant electropositive ion in the cell sap. In animals the gradient of potassium and sodium ions across the cell membrane is responsible for the potential difference across the membrane, which is important for the transmission of nerve impulses.

Potassium bicarbonate Potassium hydrogen carbonate ($KHCO_3$) A white solid prepared by passing carbon dioxide through a saturated solution of potassium carbonate. It occurs naturally in the mineral calcite.

Potassium bromide (KBr) A white solid that is extremely soluble in water. It is formed by the action of bromine on hot potassium hydroxide solution or by the neutralization of the carbonate with hydrobromic acid. Potassium bromide forms colourless cubic crystals. It is used extensively in the manufacture of photographic plate, films and papers, and as a sedative in medicine.

Potassium carbonate (pearl ash, (K_2CO_3) A white deliquescent solid manufactured from potassium chloride by the Leblanc process or, more often, using the Precht process. It cannot be made by the Solvay process. It is used in the laboratory as a drying agent and industrially in the manufacture of soft soap, hard glass and in the dyeing industry. Potassium carbonate crystallizes out between 10 and 25°C as K_2CO_3, $3H_2O$; it dehydrates at 100°C to K_2CO_3, H_2O and at 130°C to K_2CO_3. Potassium carbonate is useful in salting out alcohol from aqueous solutions.

Potassium chlorate ($KClO_3$) A white solid prepared by the electrolysis of a concentrated solution of potassium chloride. Industrially, it is prepared by the fractional crystallization of a solution containing sodium chlorate and potassium chloride. When heated, it decomposes to yield oxygen and potassium chloride. Potassium chlorate is a powerful oxidizing agent and is used in explosives, matches, weed killers, fireworks, and as a disinfectant. It oxidises iodide ions to iodine when in an acidic medium. When heated just above its melting point, potassium perchlorate is formed.

Potassium chloride (KCl) A white ionic solid prepared by neutralising hydrochloric acid with potassium hydroxide solution.

Potassium chromate (K_2CrO_4) A bright yellow solid prepared by adding potassium hydroxide solution to a solution of potassium dichromate. It is extremely soluble in water. Addition in acid to an aqueous solution of potassium chromate converts the chromate ions into dichromate ions.

Potassium cynaide (KCN) A white ionic solid that is very soluble in water and extremely poisonous. It is made industrially by the Castner process and is used as a source of cyanide and hydrocyanic acid.

Potassium dichromate ($K_2Cr_2O_7$) An orange-red solid prepared by adding potassium chloride solution to a concentrated solution of sodium dichromate and crystallizing out or by acidifying a solution of potassium chromate and evaporating. It is less soluble than sodium dichromate in cold water but more soluble in hot water. Potassium dichromate is used as an oxidizing agent both in volumetric analysis and in organic chemistry.

Potassium hydride (KH) A white ionic solid prepared by passing hydrogen over heated potassium, the metal being suspended in an inert medium. It is an excellent reducing agent.

Potassium hydroxide (caustic potash KOH) A white solid manufactured by the electrolysis of potassium chloride solution in a mercury-cathode cell. It can also be prepared by heating either potassium carbonate or potassium sulphate with slaked lime. The process closely resembles the Gossage process for making sodium hydroxide. In the laboratory, it can be made by reacting potassium, potassium monoxide or potassium superoxide with water.

Potassium iodide (KI) A white ionic solid readily soluble in water. It is prepared by dissolved iodine in hot concentrated potassium hydroxide solution. Both the iodide and iodate are formed, the latter is removed by fractional crystallization. Potassium iodide has a sodium chloride cubic lattice. In solution with iodine, it forms potassium tri-iodide. It is used in medicine, particularly in the treatment of goitre resulting from iodine deficiency.

Potassium liberated by hydrolysis from mineral The weathering of microcline, as it is conventionally written, well illustrates this formula.

Potassium losses due to leaching Much of this element is likely to be lost by leaching, sometimes to the extent of 50–70 lbs per acre.

Potassium nitrate (saltpetre, nitre, KNO_3) A white solid, soluble in water, formed by fractional crystallization of sodium nitrate and potassium chloride solutions. It occurs naturally as nitre (saltpetre) in rocks in India, South Africa and Brazil. When heated, it decomposes to give the nitrite and oxygen. This fertilizer contains 12 per cent nitrogen and 45 per cent potash in it.

Potassium nitrite (KNO_2) A creamy deliquescent solid that is readily soluble in water. It reacts with cold dilute mineral acids to produce solutions of nitrous acid. Potassium nitrite is used in organic chemistry in the process of diazotization.

Potassium nitrophos It contains 15 per cent nitrogen, 30 per cent phosphorous and 15 per cent potash.

Potassium permanganate It is used dissolved in water and has little action as a disinfectant; its antiseptic properties are due to its oxidizing powder.

Potassium phosphate It contains 30 per cent phosphorus and 30–50 per cent potash.

Potassium sulphate (K_2SO_4) A white solid prepared by the neutralization of either potassium hydroxide or potassium carbonate with dilute sulphuric acid. It occurs naturally in Stassfurt deposits as schonite. Potassium sulphate is soluble in water, forming a neutral solution. It is used as a fertilizer, particularly for tobacco plants and in the chemical industry in the preparation of alums.

Potential energy It is the energy possessed by a body due to its position. If m-mass of a body; h-its height above the surface of the earth then P.E. = mgh.

Potentiometer *See* determination of soil pH.

Potentiometric titration A titration in which an electrode is used in the reaction mixture. The end point can be found by monitoring the potential of this during the titration.

Pot experiments and fertilizer requirement Pot experiments permit test with a large number of manurial treatments within a limited space and in a relatively short time. However, as the conditions of such tests are different from those in the field, the results are not always directly applicable to large-scale farming.

Pothole A round hole worn rock streambeds by boulders, cobbles, or pebbles, rotating in eddies.

Poundal Symbol: The unit of force in the p.s. system. It is equal to 0.138 255N.

Prairie soil (brunizem) A zonal soil that develops under mid-latitude grasslands where rainfall is sufficient for leaching of soluble materials. Prairie soils are pedalfers with pH values, slightly below neutral. They form under the taller grasses that the relatively moist conditions encourage. Humus production is as a result high, giving fertility and soil structure that is very suitable for agricultural use. *See* also chernozem.

Pressure gradient The rate of change of atmospheric pressure between two points on the earth's surface. On a weather chart this is indicated by the spacing of the isobars, the gradient being 'steep' if they are close together and 'gentle' if they are far apart. This is also called barometer gradient.

Prismatic This type is characterised by vertically oriented aggregates or pillars. These elongated blocks vary in length with different soils and may reach a diameter of 6 inches. These commonly occur in the B horizon of semiarid soils.

Productiveness of a soil It is not directly proportional to the total quality of plant nutrients that it contains. It is influenced greatly by the available plant food, *i.e.*, materials which the plants are able to secure readily for their growth. The availability of nutrient ingredients is largely a matter of the various physical and biological conditions of the soil itself and the plants growing on it. Conditions conducive to decomposition of organic and inorganic components of soil and to good plant growth favour availability of plant food materials.

Products of decomposition in soil The soil organic matter contains a variety of substances at the intermediate stages of decomposition. Intermediaries products of all kinds, microbial cells, both living and dead, and stable bodies, such as waxes and lignin which already are associated with proteins, protecting them from decomposition at least for sometime and carbon dioxide is being produced in volume. Further simple products are also produced finally.

Properties of colloidal clay Depending on its nature, colloidal clay determines many of the physical as well as chemical properties of soil. Colloid material may occur as a thin gelatinous film around coarser particles, thus serving as a binding material. Colloidal clay also occur in close association with soil organic matter.

Proton An elementary particle with a positive charge (+ 1.602 192C) and rest mass $1.672\ 614 \times 10^{-27}$ kg. Protons are nucleons, found in all nuclides.

Proton number (atomic number) Symbol: Z The number of protons in the nucleus of an atom. The proton number determines the chemical properties of the element because the electron structure, which determines chemical bonding, depends on the lactrostatic attraction to the positively charged nucleus.

Protozoa in soil Three kinds of protozoa (1) amoeba (2) ciliates or infusoria and (3) flagellates. Soil protozoa are unicellular animals, feeding either or soil organic matter or on bacteria, thus regulating the number of latter in the soil. The larger forms are found in very wet condition or swamp areas.

Pumice A volcanic rock, usually derived from acid lava, which is full rounded cavities produced by expanding gases trapped in the solidifying lava. It is sometimes light enough to float on water.

Pyranose A sugar that has a six membered ring form (five carbon atoms and one oxygen atom).

Pyrenoid A protein structure found in the chloroplasts of green algae and hornworts (anthoceros). Pyrrenoids are associated with the storage of starch.

Pyrimidine A simple nitrogenous organic molecule whose ring structure is contained in the pyrimidine cytosine, thymine and uracil, which are constituents of the nucleic acids, and in thiamine (vitamin B_1).

Pyrites A mineral sulphide of a metal *e.g.* iron pyrites FeS_2, used as a soil amend.

Pyroclast A fragment of igneous rock thrown out during a volcanic explosion. The various types of pyroclast include large bombs (over 64 mm in diameter), middle-sized lapilli (including Pele's tears) and fine ash (smaller than 2 mm in diameter)

Pyrometer An instrument for measuring the temperature of molten lava.

Q

Quagmire (quacking bog) A soft bog that gives way under the feet. It may consist entirely of bog mosses or cover a stagnant pond beneath, in which case it is a floating bog.

Quantized Describing a physical quantity that can only take certain discreet values, and not a continuous range of values. Thus, in an atom or molecule the electrons around the nucleus can have certain energies, E_1, E_2, etc. and cannot have intermediate values. Similarly, in atoms and molecules, the electrons have quantized values of spin angular momentum and orbital angular momentum and orbital angular momentum.

Quantum Although light energy travels as an electromagnetic radiation embodying wave properties such as wave length and frequency unit, it is nevertheless emitted and absorbed in the form or desirable packets of energy called quanta (Sing, quantum). The energy content of light quantum is greater in the shorter wavelength of light, thus:

$$E = h \times V$$

Energy of quantum = Planks constant × Frequency of radiation

Quantum theory This is based on the idea that energy (or certain other physical quantities) can be changed only in certain discrete amounts for a given system.

Quartz The crystalline form of silica. It is most abundant of all minerals and a common constituent of many igneous, metamorphic and sedimentary rocks. Trace amounts of other substances in quartz give rise to coloured varieties such as amethyst (purple), rose quartz (pink) and cairngorm (smoky yellow or smoky brown), which all are used in jewellery. Economically also it has a great importance as it is used as gold quartz, is used as an abrasive in sand paper and toothpaste and in many other manufacturing processes including the making of pottery.

Quicksand Sand that is saturated with water and when walked on appears to liquefy; it may suck people and animals down. It is often found in river estuaries and along some coasts.

R

Rad A unit of absorbed dose of ionizing radiation, defined as being equivalent to an absorption of 10^{2-} joule of energy in one kilogram of material.

Radiant energy Solar and cosmic energy.

Radiation High-energy radiations have been effective in preserving food. Ultraviolet rays prevent the spoilage of meat because they inhibit the bacterial growth.

Radioactive Describing an element or nuclide that exhibits natural radioactivity.

Radioactive dating Any method of dating that uses the decay rates of naturally occurring radioactive isotopes to assess the age of a specimen. Organic matter less than 7000 years old can be dated using radiocarbon dating. This uses the fact that the isotope carbon-14 is found in the atmosphere and taken in by plants when they photosynthesize, and subsequently assimilated by the animals that feed on them. When plants and animals die, no more carbon is taken in and existing ^{14}C decays to the non-radioactive isotope carbon-12. If the proportion of ^{14}C to ^{12}C in the atmosphere, and the decay rate of ^{14}C to ^{12}C are both known, as they are then the sample may be dated by finding the present proportion of ^{14}C to ^{12}C. Specimens over 7000 years old can be dated by other radioisotope methods, *e.g.* potassium-argon dating.

Radioactivity The disintegration of certain unstable nuclides with emission of radiation.

Radiocarbon dating *See* carbon dating.

Radiochemistry The chemistry of radioactive isotopes of elements.

Radioimmunoassay A method for finding the concentration of peptide hormones in a biological system. Radioactive isotopes are used to label hormones which are then introduced, with specific antibodies, into the tissues or fluids that are being analysed.

Radioisotopes A radioactive isotope of an element. Tritium, for instance, is a radioisotope of hydrogen. Radioisotopes are extensively used in research as sources of radiation and as tracers in studies of chemical reactions.

Radiolysis A chemical produced by high-energy radiation (X-rays), gamma rays, or particles).

Radiometric dating *See* radioactive dating.

Radiosonde a hydrogen-filled balloon equipped with instruments for measuring and recording temperature, humidity and atmospheric pressure in the upper telemeters the information collected to a ground station.

Radio waves A form of electromagnetic radiation with wavelengths greater than a few millimetres.

Radium A white radioactive luminescent metallic element of the alkaline-earth group. It has several short-lived radioisotopes and one long-lived isotope, radium-225 (half-life 1602 years). Radium is found in uranium ores, such as the oxide pitchblende and carnotite. It is used in luminous paints and neutron sources, and (progressively less) in radiotherapy.

Symbol: ra; m.p. 700°C; b.p. 1140°C; r.d. 5(?); p.n. 88; r.a.m. 226.02.

Radon A colourless monatomic radioactive element of the rare-gas group, now thought to form compounds. It has 19 short-lived radioisotopes; the most stable, -don-222, is a decay product of radium-226 and itself disintegrates into an isotope of polonium with a half-life of 3.82 days. ^{222}Rn is used in radiotherapy.

Symbol: m.p. –71°C; b.p. –61.8°C; d. 9.73 kgm^{-1}; p.n. 86.

Rain influence on soil temperature Rain influences by (1) varying specific heat of soil, (2) by encouraging conductivity, (3) during surface evaporation and (4) by percolation. Nitrogen occurring in rain water generally is in the nitrate and ammonical forms and consequently is readily available to plants.

Rapid testing Facilities for rapid soil testing have been made available in almost all states and can be availed of. At the same time, a very large number of fertilizer experiments with different crops on a variety of soils are conducted annually in all parts of the country under a comprehensive scheme. The results of these field experiments when calibrated against those of rapid soil test will make the latter quite dependable. This will then be a valuable aid in the hand of extension workers in furnishing advice to farmers regarding fertilizer practices.

Rare elements *See* micro-nutrients. Of the elements usually found rarely in (in traces) soils and used but sparingly by plants, the application of four, under certain soil conditions and in certain crops, often give surprisingly beneficial results. These are copper, manganese, zinc and boron.

Rare gases *See* noble gases.

Ravine A deep narrow gorge, often found in arid and semiarid areas with poor vegetation cover and occasional heavy convective rainstorms. A number of gullies join together to form ravines, which may then join larger river valleys.

Ravine land reclamation The ravines have been caused due to heavy and very severe erosion over centuries and indicate an advanced stage of gully erosion. Presently many of these ravines are 60–100 feet deep with very steep vertical sides.

Raw rock phosphate It has been found in experiments that while raw rock phosphate is an excellent fertilizer, superphosphate was found to be superior

Having the formula $Ca_3(PO_4)_2$, the conventional reactions are:

$Ca_3(PO_4)_2$ + $2H_2SO_4$ = $CaH_4(PO_4)_2$ + $CaSO_4$-impurities
Insoluble Water soluble

Reactant A compound taking part in a chemical reaction.

Reaction The response that follows the application of a stimulus to a living organism.

Reaction time The time that elapses between stimulation of a whole organism and the production of a detectable response.

Reagent A compound that reacts with another (the substrate). The term is usually used for common laboratory chemicals sodium hydroxide, hydrochloride acid, etc.–used for experiment and analysis.

Reclamation of saline alkaline soils In saline-alkaline soils, the problem is two-fold; firstly, the high salt concentration and secondly poor drainage due to the presence of a hard clay pan and the highly dispersed sodium clay. In such soils, a mechanical shattering of the clay pan helps water infiltration. In soils having canker nodules in addition to the indurated clay pan, the application of gypsum accompanied by flushing with plenty of good water is necessary.

Reconnaissance This survey consists in mapping soil series and association in the scale of 1 = 1 mile using topo-sheets of the Survey of India as base maps.

Red earths The sample of red earth as a silica-alumina ratio of the clay fraction higher than 2.0 and it is fairly rich in iron oxide minerals. Its very low content of decomposable mineral hornblende indicates a highly advanced state of weathering.

Red loams The red loam samples contain a high percentage of decomposable hornblende, a silica-alumina ratio of clay fraction between 2.46 and 2.7, and a base exchange capacity below 20 m.e. per 100 gm suggesting that this soil is comparatively immature and its clay fraction is predominantly kaolinite.

Red soils The soils of this group are believed to be sedimentary formation derived from crystalline, metamorphic rocks, which consist largely of granites, genesis and schists with subordinate rocks rich in ferro-magnesium minerals.

Reduction and soil formation Under certain conditions, organisms causing decay of organic matter set up a reverse action (reduction) by taking up oxygen from simple hydroxides.

Regosol A thin azonal soil that is derived from a parent material of soft loose weathered material such as recent alluvium, loess, or dune sands. The soil is young and has no or little profile development; the A horizon directly overlies the C horizon and appears darker because of its organic content.

Regur (black cotton soil) A soil found in the Deccan Plateau of India and East Africa. Regurs are dark and alkaline as a result of a basaltic parent material. Their high clay content causes heaviness and swelling when wet and cracking when dry. Despite these disadvantages, cotton growing is widely practised on regurs.

Relative density Symbol: d The ratio of the density of a given substance to the density of mass reference substance. The relative densities of liquids are usually measured with reference to the density of water at 4°C. Relative densities are also specified for gases; usually with respect to air at STP. The temperature of the substance is stated or is understood to be 20°C. Relative density was formerly called specific gravity.

Relative molecular mass Symbol: M_1, The ratio of the average mass per molecule of the naturally occurring form an element or compound to 1/12 of the mass of an atom of nuclide ^{12}C. This was formerly called molecular weight. It does not have to be used only for compounds that have discrete molecules; for ionic compounds (*e.g.* NaCl) and giant molecular structures (*e.g.* BN) the formula unit is used.

Relative velocity The relative velocity of a body A with respect to B can be found by compounding a velocity equal and opposite to that of B with the velocity of A.

Residual soil It is formed *in situ* from underlying rocks, geologically old.

Residual soil material This kind of material is developed after suffering a long and intense gathering. If occurring in a warm climate, it is roughly oxidised and well leached. It may be comparatively low in calcium even though having originated from limestone.

Resolving power (resolution) The ability of an optical system to form separate images of closely spaced objects. *See* microscope.

Retention of P_2O_5 in soil The phosphoric acid in superphosphate is wholly water soluble, but when applied to soil, it is immediately converted into insoluble phosphates due to precipitation as calcium, iron or aluminium phosphate, according to as the soil is alkaline or acid. Thus the fertilizer is not leached but is slowly dissolved in the soil solution. Fixation losses can be reduced by applying the fertilizer in bands on either side of four to six inches by means of a drill. By this means, at least some of the phosphate does not come in direct contact with the soil and thus the available phosphorus is readily released for absorption by the plant roots. The fertilizer is suitable for all crops and can be applied to all soils. In acid soils, it should properly be used in conjunction with organic manure. It should be applied before or at sowing or transplanting.

Reversible reaction A chemical reaction that can proceed in either direction. For example, the reaction:

$$N_2 + 3H_2 = 2NH_3$$

is reversible. In general, there will be an equilibrium mixture of reaction and products.

Reverted phosphoric acid This acid is never added in amounts capable of completing this reaction. Consequently some calcium-phosphate $Ca_3H_2(PO_4)_2$ spoken of as citrate soluble is produced:

$$Ca_3\ (PP_4)_2 + H_2SO_4 = Ca_3H_2\ (PO_4)_2 + CaSO_4 + \text{impurities}$$

Richter scale A scale that indicates the quantity of energy released by an earthquake (*i.e.* the magnitude of the earthquake), devised by the US seismologist Charles F. Richter in 1935. The scale ranges from 0 to 9, but in practice there is no upper limit. The largest observed magnitude was 8.6.

Rigolith (waste mantle) A layer of unconsolidated weathered material lying above bedrock and below the ground surface. It includes the parent material from which soil is formed and the soil itself. It consists of weathered particles of the bedrock itself or mineral particles transported into the area by the action of water, wind, or ice.

Rill erosion Erosion of the soil surface by shallow short-lived channels known as rills. Rills form during storms when water flowing over the surface of the soil is deep enough to become turbulent

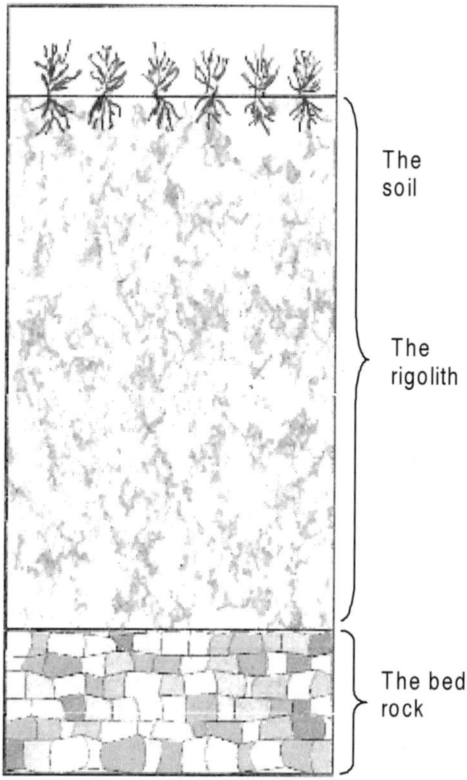

The Position of Rigolith

and erosive or when water flowing underground re-emerges on lower hill slopes and washes away the soil. If the rainfall continues, the rills may be further eroded and enlarged to form gullies.

River water solutes The material in solution usually consists of various salts of which sulphates, chlorides, carbonates, bicarbonates, and sometimes nitrates of calcium, magnesium, sodium and sometimes potassium, are the most common. Some waters may also contain sulphur, iron or even some minor or trace elements, *e.g.* boron, fluorides, selenium, etc. Waters differ not only in the kind of salts they contain but also in the total salt contents. The nature and quantity of salts present in water affects its suitability for irrigation. Depending upon the amount and nature of salts it contains, a water may be suitable or unsuitable for irrigation.

Rock Any aggregate of mineral particles that forms part of earth's crust. It may be unconsolidated as clay, mud and sand. *See* classification of rocks. Three major groups are found–igneous, metamorphic and sedimentary.

Rock phosphate It occurs as natural deposits of rock in many places in the world.

Rock salt The crystalline form of common salt (sodium chloride, NaCl). It occurs as an evaporite deposit as a result of evaporation of enclosed or partially enclosed bodies of sea water or salt lakes and is often found in association with other mineral salts. Rock salt is extracted either by mining or by pumping brine to the surface where it is recorded by evaporation.

Rotifers These occur in soil and under moist conditions, especially in swamps, their numbers may be big enough. Many species are found living in swampy conditions.

Runoff The portion of rainwater melt-water from snow that drains into rivers as opposed to evaporating or being lost through transpiration by plants. Other runoff are direct runoff, surface runoff and delayed runoff.

Runoff coefficients Very greatly due to differences in slope, cover, soil material and state of cultivation. The values of 'C' as suggested by Ramser of American conditions are given in the table below:

		Percentage Slope		
Land cover	1–5	5–10	10–30	30–40
		Runoff Coefficient		
Timber	0.20	0.20	0.20	0.30
Pasture	0.30	0.35	0.40	0.60
Cultivated	0.50	0.60	0.70	1.00

The rational for calculating runoff is based on the principle that the maximum runoff will occur when a rainfall of maximum intensity continues for a time equal to or exceeding the time of concentration, *i.e.*, T, the latter being the interval of time required for water to flow from the most distant point in the water-shed to the point of concentration structure. The 'time of concentration' for any catchment area is calculated by dividing the length of slope by the velocity of the runoff of that slope. The above table shows the approximate average velocity of runoff for different slopes under varying land covers.

S

Saccharide *See* sugar.

Sachse reaction A process for the manufacture of ethane from natural gas (methane). Part of the methane is burned in two stages to raise the furnace temperature to about 1500°C. Under these conditions, the rest of the methane is converted to ethane and hydrogen:

$$2CH_4 = C_2H_2 + 3H_2$$

The process is important since it provides a source of ethane from readily available natural gas, thus avoiding the expensive carbide process.

Sal ammoniac *See* ammonium chloride.

Salina A very salty or dried up playa lake at the centre of a basin of inland drainage in an arid region, chiefly in SW North America. *See* also play a (Spanish).

Saline Containing a salt, especially in alkali-metal halide such as sodium chloride.

Saline soils A group of intrazonal soils that contain high concentrations of salts such as common salt (NaCl). They often occur in semiarid and arid areas where there is strong evaporation and where the parent material due to ground evaporation and where ground water contains salt. Near salt lakes, saline soils merge with pure salt crusts. Saline soils pose problems for irrigation. *See* also solonchak, solonetz. Saline soils have a permeable sub-soil and a high water table at some part of the year. They show extensive white, greyish white or ash coloured fluffy salt deposits on the surface in the dry season.

Salinity The saltiness of water, usually expressed in parts per thousand. This varies greatly throughout the parts of the world, especially in the surface layers. The main determining factors are the rate of evaporation and the rate of influx of fresh water (rivers and precipitation).

Salinization The process by which soils are enriched with salt. Salinization may occur as a result of evaporation from the soil surface, which draws up salts in solution by capillary movement. It may also be induced by man, for example, irrigation practised where evaporation is intense flushing, deep ploughing, or chemical treatment. *See* also saline soils. In low rainfall, arid or semi-arid regions, the soluble salts tend to accumulate either in situ in the soil if drainage is faulty, or movement in ground water to low lying places where seepage accumulates.

Salt A compound with an acidic and a basic radical, or a compound formed by total or partial replacement of the hydrogen in an acid by a metal. In general terms a salt is a material that has identifiable cationic anionic components.

Salt, common, *See* rock salt.

Saltation The major portion of the soil carried by wind is moved in a series of short bounces called saltation. The soil carried in saltation consists of the particles ranging from 0.1 to 0.15 mm in diameter. Saltation is caused by the direct pressure of the wind on soil particles. The initial rise of a grain in saltation is generally in a vertical direction.

Salt hydrate A metallic salt in which the metal ions are surrounded by a fixed number of water molecules. The water molecules are bound to the metal ions by ion-dipole inter-actions. In the solid state, the water molecules form part of the crystal structure. Salt hydrates that contain several molecules of water for each metal ion can lose them progressively (effloresce) if the dissociation pressure of water vapour is kept below the dissociation pressure of the system; eventually an anhydrous salt is formed. Alternatively, if the vapour pressure is continuously raised, the system will add molecules of water (deliquesce) until a saturated solution forms.

Saltpetre *See* potassium nitrate.

Sand Particles of rock of a size intermediate between silt and gravel; internationally defined as particles with diameters ranging from 0.06 to 2.00 mm. Fine sand ranges up to 0.2 mm. Most sands are mainly composed of the hard-wearing mineral quartz.

Sandstone A clastic sedimentary rock consisting primarily of grains of quartz, together with grains of mica, feldspar, and other minerals in small quantities. The sand accumulates through the action of wind or water and is laid down in shallow seas, estuaries, deltas, lowlying coasts, and deserts. Sandstones may be classified calcareous, siliceous, or ferruginous according to the nature of the material cementing the grains together.

Sandy soils These are easy to work and are, hence, called light soils. Workability of, on the other hand, clay soils are difficult to till.

Saponification The process of hydrolysing an ester with a hydroxide. The carboxylic acid from a sodium salt. For instance, with sodium hydroxide; RCCOR′ + NaOH – NaOOCR + R′OH. The saponification of fats, which are esters of 1,2,3-trihydroxy-propane with long-chain carboxylic acids, forms soaps.

Saturated compound An organic compound that does not contain any double or triple bonds in its structure. A saturated compound will undergo substitution reactions but not addition reactions, since each atom in the structure will already have formed its maximum possible number of single bonds.

Saturation In meteorology, the state of a mass of air, at particular temperature, when it contains as much water vapour as it can possibly hold *i.e.* as many molecules of water enter the air as many leave it. The amount of water vapour required to saturate the air depends on temperature and pressure; the amount increases as temperature increases. Condensation in the atmosphere occurs when saturated air is cooled, thus losing the capacity to retain all the water vapour within it.

Saturated solution A solution that contains the maximum equilibrium amount of solute at a given temperature. A solution is saturated if it is in equilibrium with its solute. If a saturated solution of a solid is cooled slowly, the solid may stay temporarily in solution; *i.e.* the solution may contain more than the equilibrium amount of solute. Such solutions are said to be supersaturated.

Saturated vapour A vapour that is in equilibrium with the solid or liquid. A saturated vapour is at the maximum pressure (the saturated vapour pressure) at a given temperature. If the temperature of a saturated vapour is lowered, the vapour condenses.

Scale of pH value A pH value of 7.0 denotes a neutral solution, containing 0.0000001 gm. of hydrogen ions (H^+), and equivalent amount of hydroxyl (HO)-ions in one litre of water. As the soil becomes increasingly acidic, the pH value progressively decreases below 7.0. The scale given below shows the relation between pH values and the degree of acidity:

pH	*Reaction*
7.0	Neutral
7.0 to 6.0	Slightly acidic
6.0 to 5.5	Medium acidic
5.5 to 5.0	Strongly acidic
5.0 to 4.5	Very strongly acidic
4.5 to 4.5	Extremely acidic

Screw gauge This instrument is used for measuring the diameter of wires and thickness of thin sheets. It contains a screw which rotates in a linear or pitch scale. There is a circular or cap scale on the screw.

Second law of motion The rate of change of momentum of a body is proportional to the impressed force and it takes place in the direction of the force.

Second's pendulum It is a simple pendulum whose time period is 2 secs.

Sedentary *See* transport of soil material.

Sediment Particles deposited by a fluid. Sediments are most commonly composed of rock particles, but organic sediments or mixed organic and mineral sediments also occur. Aeolian sediments are those deposited by the wind; fluvial, marine, and fluvioglacial sediments result from deposition by water. Sediments range in size from clays (less than 0.002 mm in diameter), through silt, sand, gravel, cobbles and boulders.

Sedentary soil material When unconsolidated debris might lie in position above bedrock for centuries.

Sedimentary rock A rock composed of sediments and generally having a layered (stratified) appearance. Sedimentary rocks from one of the three major groups of rocks that together make up the earth's crust (the other two groups being igneous and metamorphic rocks). The sediments are largely derived from pre-existing rocks that have been broken down and then transported by water, wind, or ice.

Sediment Particles deposited by a fluid. Sediments are most commonly composed of rock particles but organic sediments or mixed organic and mineral sediments also occur.

Sediment yield The amount of sediment removed by rivers from their drainage basin over a period of time, usually in tonnes per year.

Seepage The movement of water through the pores of the soil sediments or porous rocks.

Seif dune (longitudinal dune) A desert sand dune in the form of a long ridge, parallel to the prevailing wind.

Seliwanoff's test A standard test for the presence of fructose in solution. A few drops of Seliwanoff's reagent resorcinol in hydrochloric acid are heated with the test solution. A red colour or red precipitate indicates fructose.

Series In geology, the rock layers deposited during one epoch.

Serozem *See* sierozem.

Sewage sick soil The soil quickly becomes 'sewage sick' due to use of sewage without treatment makes it sick.

Shale An argillaceous sedimentary rock composed of layers of clay and silt particles with a grain size of 1–16 mm to 1–10^{24} mm in diameter. It is one of the most abundant sedimentary rocks.

Sheet erosion It removes a thin covering of soil from large areas, often entire fields, more or less uniformly during every rain producing a runoff. This type is very insidious which keeps the cultivator almost ignorant of its ill-effects (sheet, wash, rain wash). Erosion by water flowing over a relatively smooth slope is important in sediment formation.

Sheet flow (sheetflood) The flow of water over the surface of the land in the form of thin continuous sheets. Very shallow sheetflow may not result in erosion, but deeper flows are turbulent and erosive causing sheet erosion, particularly where vegetation cover is poor.

Shingle A deposit of coarse gravel especially on a beach.

Sial The relatively light socks that forms the continental crust. These are composed predominantly of silica and aluminium, *i.e.* acid rocks such as granite. *See* also sima (silica aluminium).

Sierozem (serozem, grey desert soil) A soil formed in cool or temperate arid areas under scrub vegetation, the litter of which is sparse; the overall grey colour of the soil reflects this paucity of organic matter. Calcium carbonate accumulates at or just below the surface, sometimes as concentrated pan, and the whole soil profile is alkaline.

Silica The mineral silicon dioxide (SiO_2) the most abundant of all minerals on earth's crust. It occurs naturally in at least nine different forms.

Silicate minerals A group of minerals that constitutes about one third of all minerals and consists of silicates of calcium, magnesium, aluminium and other metals. All are based on a structure unit known as the SiO, tetrahedron. The silicate minerals include the feldspars, mica, clays, talc, and beryl. Many are economically important.

Silicate clay base exchange Evolution of CO_2 and carbonic acid after the decomposition of organic matter, the ionic hydrogen tends to be exceedingly active and will tend to replace the exchangeable calcium of the colloidal complex. This occurs not only due to mass action effect but also because ionic hydrogen is adsorbed more strongly than is the ionic calcium.

Silicate clays mineralogical nature For many years even after its colloidal nature was recognised, clay was thought to be an amorphous body. X-ray examinations show definitely that in spite of their much smallness, the lamellar clay particles are crystalline.

Silicate clay molecular make up Each clay particle regardless of its individual shape is made up of sheet like molecules or units, held loosely together. These units show great variations among themselves at least in size.

Silicon The second most abundant element in the earth's crust, after oxygen. It forms a major constituent of nearly all rock-forming minerals. Pure silicon is of major use in the electronics industry as a semiconductor. It is found in soils as SiO_2 and is present in various amounts ranging from 70 to 91 percent.

Silt Fine grains of mineral material in size sand clay with diameters from 0.06 to 0.002 mm.

Silver A greyish-white metallic element. It occurs in nature as the metal, frequently in association with other metals such as copper and gold, as argentite (Ag_2S), and in association with lead, zinc and copper ores. Silver is readily workable and is used in coinage, sliver plate and jewellery. Pure silver has the highest thermal and electrical conductivity of all the metals and is used in some printed electrical circuits.

Silver (I) bromide (AgBr) A pale yellow precipitate obtained by adding a soluble bromide solution to a solution of silver (I) nitrate. It is used extensively in photography for making the sensitive emulsions for plates and films.

Silver (I) chloride (AgCl) A compound prepared as a curdy white precipitate by the addition of a soluble chloride solution to a solution of silver (l) nitrate. It occurs in nature as the mineral horn silver.

Silver (I) iodide (Agl) A pale yellow precipitate prepared by the addition of a soluble iodide solution to a solution of silver (l) nitrate. Silver (l) iodide occurs in nature as iodoargyrite in the form of hexagonal crystals.

Silver-mirror test A test for the aldehyde group. A few drops of the sample are warmed with "Tollen reagent". An aldehyde reduces Ag^+ to silver metal, causing a brilliant silver mirror to coast the inside wall of the tube.

Simple pendulum It is a small heavy particle suspended by an elastic and weightless string. If l is length of simple pendulum (=distance between the poles of suspension and *e.g.* of the bob) then: $t = 2\Pi\sqrt{1/g}$ or $g = 4\Pi^2 1/t^2$.

Single bond A covalent bond between two elements that involves one pair of electrons only. Often referred to as a sigma bond, (a bond), it is represented by a single line, for example H-Br. The sigma bond is formed by the overlap of sigma-atomic orbitals from each atom (*i.e.* orbitals that are on the inter-nuclear axis).

Skeletal soil *See* azonal soil.

Slaked lime *See* calcium hydroxide.

Slaking The stresses caused in the rock by alternate weathering and drying may lead to its disintegration. Hydration is particularly effective in clays where it causes the break up or slaking of the clay.

Slate A fine-grained metamorphic rock in which the arrangement of flaky minerals such as mica into parallel layers gives the rock its characteristics structure known as flaky cleavage. Slates are formed by the metamorphism of mudstones, silistones, and other fine-grained deposits.

Slump A mass movement of rock debris or soil, often lubricated by water that rotates backwards as it slips on a curved concave surface. It leaves a scar on the hill-slope from which it has sheared.

Slurry A thin paste of suspended solid particles in a liquid.

Soap One of a number of sodium or potassium compounds of certain organic acids that are commonly used to improve the cleaning properties of water. Soap was the earliest known detergent. It is made first reacting vegetable oils and animal fats with a strong solution of sodium or potassium hydroxide to produce organic acids, such as octadecanoic acid.

Soda ash *See* sodium carbonate.

Soda lime A grey solid produced by adding sodium hydroxide solution to calcium oxide, to give a mixture of $Ca(OH)_2$ and NaOH. It is used in the laboratory as a drying agent and to absorb carbon dioxide.

Sodamide A white ionic solid formed by bypassing dry ammonia over sodium at 300 to 400°C. The compound reacts with water to give sodium hydroxide and ammonia. It reacts with redhot carbon to form sodium cyanide and with nitrogen (I) oxide to form sodium acid. Sodamide is used in the Castner process and in the explosives industry.

Sodium It is not an essential element for plant growth, but some crops such as beet, celery, cabbage, kale, knoi-khol, radish, rape and turnip benefit greatly by the application of soluble sodium salts; especially if the soil is deficient in potassium. Sodium is also of direct benefit to plants indigenous to seashore or to irrigated arid regions.

Sodium absorption by colloids in soils Nitrate of soda supplies nitrogen in a form that immediately stimulates the crop even if the soil is cold. It tends to reduce the acidity of soils slowly or even tends to produce sodium carbonate which is quite saline.

Sodium aluminate ($NaAlO_2$) A white solid produced by adding excess aluminium to a hot concentrated solution of sodium hydroxide.

Sodium adsorption by colloids *See* nitrate of soda.

Sodium in acidic soils The presence of as much sodium as the chloride, carbonate and sulphate tends more or less to saturate the colloidal matter of typical alkali soils with this element resulting in sodium complex.

Sodium acid (NaN_3O_3) A white solid prepared by passing nitrogen (l) oxide over heated sodamide. On heating, sodium acid decomposes to give sodium and nitrogen. It is used as a reagent in organic chemistry and in the preparation of lead acid (for use in detonators).

Sodium benzoate *See* sodium benzenecarboxylate. It is also used as a food preservative.

Sodium carbonate (soda ash Na_2CO_3) A white amorphous powder, which aggregates on exposure to air owning to the formation of hydrates. Industrially it is prepared by the ammonia-soda (Solvay) process.

Sodium chloride (common salt, salt, NaCl) A white solid prepared by the neutralization of hydrochloric acid with aqueous sodium hydroxide. It occurs in seawater and natural brines. Natural solid deposits of rock salt are also found in certain places.

Sodium chloride structure A form of crystal structure that consists of a face-centred cubic arrangement of sodium ions with chloride ions situated at the middle of each edge and in the enter of the cube.

Sodium cyanide (NaCN) A white solid formed either by the action of carbon on sodamide at high temperature or by passing ammonia over sodium at 330–400°C to form sodamide and reacting it with carbon.

Sodium dichromate ($Na_2Cr_2O^-$) An orange deliquescent solid that is very soluble in water. It is made from finely powdered chromite, which is heated with calcium oxide and sodium carbonate in a furnace. If an alkali is added to a solution of the dichromate, the chromate is produced. At high temperatures, sodium dichromate decomposes to give the chromate, chromic oxide and oxygen.

Sodium ethanoate (sodium acetate, ($CH_3COOHNa$) A white solid prepared by the neutralization of ethanoic acid with either sodium carbonate or sodium hydroxide.

Sodium exchange in soils When the drainage of an aridic soil is impeded and soluble salts accumulate, Na ions are likely to equal or even exceed those of absorbed Ca.

Sodium fluroide (NaF) A white solid formed by the action of hydrogluroric acid on sodium carbonate or sodium hydroxide (the reaction between metallic sodium and fluorine is too violent).

Sodium flurosilicate It has been used as an insecticide from time to time and has certain advantages. It is cheaper than the arsenicals and kills insects more rapidly.

Sodium hexafluoroaluminate (cryolite, Na_3AlF_6) It is white or colourless, but may be reddish or brown due to impurities. Sodium hexafluoroaluminate is used as a fluxion in the manufacture of aluminium.

Sodium hydride (NaH) A white crystalline solid prepared by passing a pure stream of dry hydrogen over sodium at 350°C; the sodium is usually suspended in an inert medium. Electrolysis of fused sodium hydride yields hydrogen at the anode. Sodium hydride reacts with water to form sodium hydroxide solution and hydrogen.

Sodium hydroxide Caustic soda (NaOH) A white deliquescent slightly translucent solid with a fibrous texture. In solution, it is a strong alkali and electrolyte. In the laboratory, it can be prepared by reacting sodium, sodium monoxide or sodium peroxide with water.

Sodium iodide (NaI) A white solid formed by reacting sodium carbonate or sodium hydroxide with hydroiodic acid; the solution is then evaporated. Sodium iodide forms colourless crystals having a cubic shape. It is used as a source of iodine and in medicine and photography. Acidified solutions of sodium iodide exhibit reducing properties owing to the formation of hydroiodic acid.

Sodium methanoate Sodium formate (HCOONA) A white solid produced by reacting carbon monoxide with solid sodium hydroxide at 200°C and 10 atmospheric pressure. Alternatively it can be prepared by the neutralization of methanoic acid with sodium hydroxide solution.

Sodium monoxide (Na_2O) A white solid formed by burning sodium in a deficiency of oxygen or alternatively of reducing sodium peroxide or sodium hydroxide with the requisite amount of sodium. It reacts violently with water to form sodium hydroxide and with acids to form solutions of their salts.

Sodium nitrate (Chile saltpetre, $NaNO_3$) A white solid formed by the neutralization of nitric acid with either sodium carbonate or sodium hydroxide. Sodium nitrate is very soluble in water and on crystallization forms colourless deliquescent crystals. On heating, it decomposes to give sodium nitrite and oxygen. It is a chemical fertilizer, $NaNO_3$. *See* nitrate of soda.

Sodium nitrite ($NaNO_2$) A yellowish-white solid formed by the thermal decomposition of sodium nitrate. It is readily soluble in water. The anhydrous salt forms orthorhombic crystals. When treated with cold dilute hydrochloric acid, sodium nitrite forms nitrous acid. It is used in organic chemistry in the process of diazoization and industrially as a corrosion inhibitor.

Sodium peroxide (Na_2O_2) A yellowish-white ionic solid formed by the direct combination of burning sodium with excess oxygen. It reacts with water to form sodium hydroxide and hydrogen peroxide; the latter decomposes rapidly in the alkaline solution to give oxygen. Sodium peroxide can covert nitrogen (II) oxide to sodium nitrate and iodine to sodium iodate.

Sodium sulphate (Na_2SO_4) A white soluble solid formed by heating a mixture of sodium chloride and concentrated sulphuric acid. Sodium decahydrate, known as Glauber's salt, is used in the manufacture of glass and as a purgative in medicine.

Sodium superoxide (sodium dioxide, NaO_2) A pale yellow solid obtained by heating sodium peroxide in oxygen at 490°C. It reacts with water to give hydrogen peroxide, sodium hydroxide solution and oxygen. Commercial sodium peroxide contains 10 per cent sodium superoxide.

Sodium thiosulphate ($Na_2S_2O_3$) A white solid prepared either by boiling sodium sulphite with flowers of sulphur or by passing sulphur dioxide into a suspension of sulphur in boiling sodium hydroxide.

Soft soap A liquid soap made by saponification with potassium hydroxide (rather than sodium hydroxide).

Soft water *See* hardness.

Soil The accumulation of mineral particles and organic matter that forms a superficial layer over large parts of the earth's surface. It provides support and nutrients for plants and is inhabited by numerous and various micro-organisms and animals. A section down through the soil is termed a soil profile and this can characteristically be divided into three main layers or horizons. Horizon B, the subsoil, contains materials washed down from above and may be mottled with various colours depending on the iron compounds present. Horizon C is relatively unweathered parent material from which the mineral components of the above layers are derived. The depth and content of the horizons are used to classify soils into various types, *e.g.* podsols and brown earths. The texture, structure and porosit of soil depend largely on the sizes of the mineral particles it contains and on the amount of organic material present. Soils also very depending on environmental conditions, notably rainfall. *See* also humus.

Soil acidity increase This can be done by adding more and more organic matter, addition of chemicals such as ferrous sulphate and flowers of sulphur etc.

Soil's active portion Clay and humus, intimately mixed, constitute most of the colloidal matter found in soils. This finely divided portion not only promotes activity catalytically, but also may participate intensively in the changes that occur.

Soil adaptability To define their adaptability for crops, grasses and trees, their management requirements and yields of individual crops under different systems of land use and management.

Soil aeration and plant growth Furthermore, due to their open nature, the soils are well aerated, whereas the clay soils absorb water slowly and hence are liable to water-logging and the resultant aeration or anaerobic conditions. Sandy soils do not do so.

Soil air Deep ploughing, sub-soiling and cracking of soils in summer tend to improve soil aeration. It has been seen that 30 to 66 per cent of the soil volume is pore space, and the pore space not filled with water is occupied by air. The soil air is not a continuous gaseous body as the lower atmosphere. But it is highly dispersed and adsorbed by soil colloids. Its humidity usually approaches 100 per cent, a condition especially favourable for the growth of fungi and other soil organisms.

Soil amendments Lime is generally used for correcting soil acidity, improving physical condition of soil and encouraging bacterial activity. Similarly, gypsum is used for the reclamation of 'alkali' soils or land from the sea and improving the structure of heavy black clay soils. Hence, these are called soil amendments.

Soil analysis Laboratory testing of soil is generally supplementary to field examination and has usually two objectives: 1. To provide a basis for classification of soils and comparison of their origin and constitution, and 2. To determine the plant nutrient contents of soil so that deficiencies, if any, may be made good by suitable manurial treatments.

Analysis of Different Kinds of Soils

100% sand	100% clay	100% silt
80% sand and 20% clay	50% clay and 50% silt	20% silt and 80% sand
Sandy–Clay	Silty–Clay	Loam–Sandy
70% sand and 50% clay	30% clay and 50% silt	50% silt and 50% sand or clay
Sandy–Clay–Loam	Silty–Clay	Clay–Loam or Silt–Loam or Silt–Loam–Sandy
50% sand and 30% clay	20% clay and 50% silt	50% silt
Sandy–Clay	Silty–Clay–Loam	Loam

Soil blue grass A good binding plant material used to prevent soil erosion of the grazing lands.

Soil biology A soil bearing as it does so much available energy in organic form, harbours a varied population of living organisms.

Soil buffering In affecting a definite change of pH in a soil it is usually necessary to have much more acid or alkali than the amounts of H and OH ions in the soil solution would lead one to expect. In other words there is a distinct resistance to a change in pH other than that offered by soil solution itself. This is called buffering and is due to the weak acids and their salts.

Soil's capillary conductivity Capillary movement of water depends much on the distance through which the adjustment may take place. Though distant is also important but the prime consideration is how much water can be delivered at any particular point in unit time. This is what is known as capillary conductivity.

Soil class determination The common method of class determination is the one employed in the field. It consists in an estimation organic content, and especially in a testing of the feel of the soil in order to decide as to the class name.

Soil classification As it applies to various individual situation, the process is disruptive as well as destructive. The formation of clay and the oxidation of iron are the good examples of synthesis and stabilization of the weathering processes and thus the regolith is slowly prepared. Also *See* classes of soil.

Soil classification or grouping On the basis of nature of the soil forming environmental factors, the soils are broadly grouped.

Soil colloidal system It is a complex in which a tremendously large interface is attained by the very fine, heterogeneous dispersion of one or more substances in a second substance or material.

Soil condition for growth of fungi They respond especially to soil aeration, their numbers and activities diminishing as air movement is retarded. Moulds develop vigorously in acid, neutral and alkaline soils but they are abundant in acid soils where bacteria and actinomyces offer a least competition.

Soil condition and availability of nutrients Much of nitrogen is utilized by higher plants is absorbed in nitrate and ammonium forms, the importance of these processes is obvious. Their transformation takes place during bacterial and fungi in the soil, which are profoundly influenced by soil condition. When the soil is cold or water-logged, these bio-chemical changes do not progress rapidly. Again the lack of active lime, allowing the soil to become acid, retards these processes.

Soil conservation The basic factor in the application of erosion control and water conservation measure is the proper utilisation of the land according to its capabilities. Broadly speaking, the practical methods of soil conservation fall into two important classes: biological and mechanical. The biological measures may again be divided into (*i*) agronomic (*ii*) dry farming and (*iii*) agrostological. Each of these conservation techniques should supplement and strengthen the others in an integrated programme.

Soil conservation and grasses Some important grasses and legumes that have shown promise for soil conservation are *Cynodon dactylon, Cynodon plectistachyium*, for stabilization of new embankments; *Dichanthium annulatum, Panicum antidotale*, for consolidating field bunds.

Soil conservation by legumes Amongst this type of plants, the promising species are *Peuraria hirsute, Pueraria phaseoloides, Dolichos lablab* and *Indigofera endecaphylla*.

Soil content of fertilizer elements It is difficult to generalise on this subject because the soils of different localities, condition of soil (whether cultivated or barren) and the cultivation practices mainly determine the soil content of particular locality. For such types of studies, a soil analysis may be necessary.

Soil density Soils having larger particles are usually heavier in weight per unit volume than those with smaller particles. True density of soil is based on the individual densities of soil constituents and according to their proportionate contribution.

Soil depletion of nutrients By cultivation practices.

Treatments	*Depletion of Nutrients in Kg/Hectare*					
	N	*P_2O_5*	*K_2O*	*CaO*	*MgO*	*SO_2*
Field in which maize every year	85	50	910	385	195	51
Field in which maize wheat and clover crop rotation was practices	31	20	320	150	60	20

Soil effected by lime Two effects are observed by addition of lime to the soil, first is physical and second is chemical. In the first, granulation takes place. Chemically lime brings about many complex changes in an acid soil by reducing the H-ion concentration and tends to force such cations as aluminium, iron and manganese into an insoluble condition.

Soil earth worms These are important creatures of soil. The hole left in soil serve to increase erosion of soil and drainage.

Soils' edaphalogical definition It is a natural body engendered from a variable mixture of broken and weathered minerals and decaying organic matter, which covers

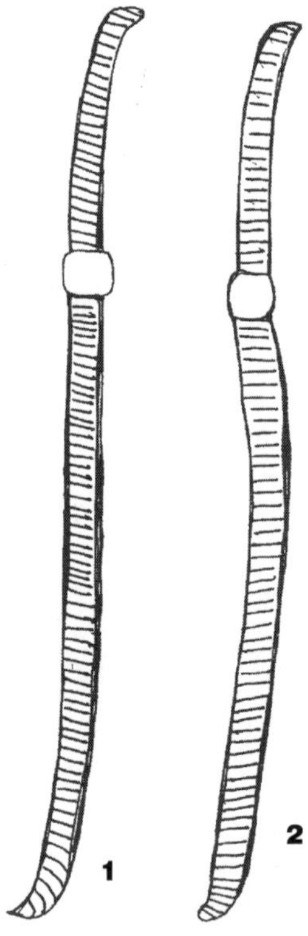

Two of the Common Earthworms Present in Arable Soils

1. *Lumbrius terrestris;*
2. *Allolobophora caligrinaso*

the earth in a thin layer and which may supply, when containing proper amounts of air and water, mechanical support and in part substance for plants.

Soil erosion The removal of soil, either mechanically or chemically. Wind and water are the main agents of soil erosion. Water may cause surface erosion by overland flow in the form of sheet erosion, rilling and gullying.

Soil erosion and its ill effects It works out its havoc through loss of surface solid, plant food, sub-soil water, silting up of water channels, disruption of communications, reduction of productivity of land and ultimately lowering the already low standard of living of the peasantry.

Soil-energy relation with water The manner in which water moves or is held in the soil, its relationships to plants, and to what extent its translocations and amounts may be controlled must be reckoned within any study of soil and plant relationships.

Soil's exchangeable and actual acidity In an acid soil, two more or less distinct sets of hydrogen ions are involved those or the soil solution itself and those held adsorbed cations by the colloidal complex. These groups tend towards an equilibrium, are mutually supporting, and must both be considered in any attempt to alter the H-OH ration of the soil solution. For distinguishing the H-ion concentration of the soil solution, it is designated as active or actual acidity and the H-ions held in abeyance, or in reserve are referred to as the exchangeable or potential acidity of the soil.

Soil fertility No two soils are alike in the nature or the quantities or plant nutrients they contain. In a given situation, their system of farming, soil management and manuring practices, etc. influence the productivity of soils and crop yields.

Soil fertility and magnesium As most limes contain considerable magnesium, dolomite limestone sometimes carry 40 per cent of magnesium carbonate.

Soil fertility management in phases The principal contributors to the successful maintenance of soil fertility are (1) Adequate supply of nitrogen (2) Addition of soil organic matter by various means (3) Application of lime is necessary (4) Provision of phosphorus by various means and (5) Application of trace elements.

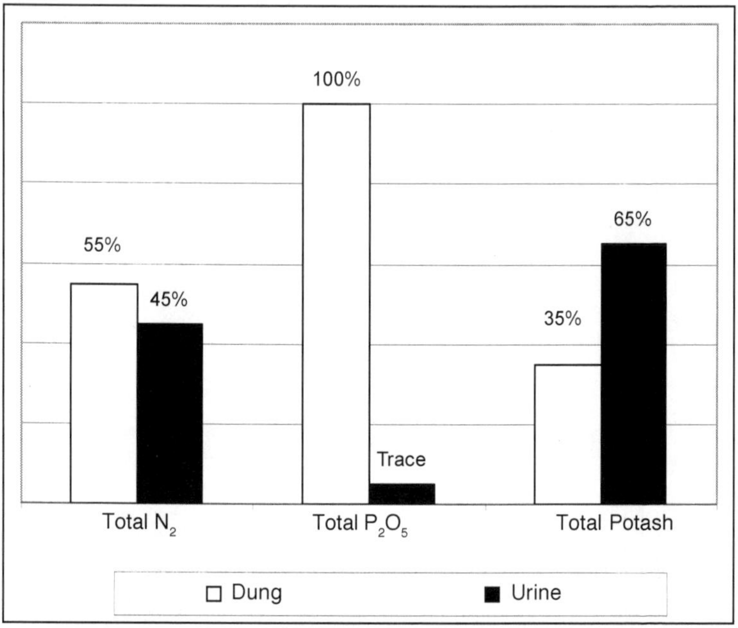

Distribution of Nitrogen, Phosphoric Acid and Potash in the Solid and Liquid Portions of an Average Farm Manure

Soil formation Practical knowledge of crop relationship in any region is of uncertain application without some understanding of the formative processes of the soils in question and of their present profile similarities and differences. For this, three phases should be considered (1) evolution of soil from parent material (genesis) (2) soil classification and (3) soil survey.

Soil forming rocks Some more common rocks are:

Igneous	*Sedimentary*	*Metamorphic*
Granite	Limestone	Genesis
Syenite	Dolomite	Schist
Diorite	Sandstone	Slate
Gabbro	Shale	Quartzite
Periodorite	Conglomerate	Marble

Soil fumigants These materials are employed for killing all plant growth before seeding to more desirable species. They function as a vapour or gas that diffuses through the soil and have a relatively short life in the soil as a consequence of which replacing in the treated area is possible within four weeks or less. Carbon disulphide, chloropicrin, vapam, methylbromide and cyanamide are some soil fumigants in use.

Soil groups according to formation Fresh alluvium and newly formed dry sands or stony soils on steep hill sides are examples of the azonal group. The soils may also be grouped accordingly as their formation takes place immediately over the underlying parent rock.

Soil heat A certain degree of heat is necessary for seed germination, root growth and microbiological activity in the soil. All these developmental processes almost cease at or near the freezing point of water, but the soil temperature seldom falls so low except at the surface in the cold season in some parts of North India or high up in the mountainous regions.

Soil horizon A well-defined layer within the soil profile, parallel to the local ground surface. Three main horizons–A, Band C are characteristics of many free-draining soils. Such horizons are visually distinctive, reflecting their different physical and chemical properties, which result from soil-forming processes including weathering, the introduction of humus, and movement of minerals. *See* also A horizon, B horizon, C horizon and soil profile.

Soil humus composition The average composition of humus in terms of the end products is as follows:

Carbon	50 per cent
Oxygen	35 per cent
Nitrogen	5 per cent
Hydrogen	5 per cent
Ash (containing phosphorus, potassium, sulphur and many other elements)	5 per cent

Soil-inorganic colloids The clayey matter of soil is the seat of such varied and vigorous activities that it deserves a thorough study. In fact, every one feels that many problems, especially those

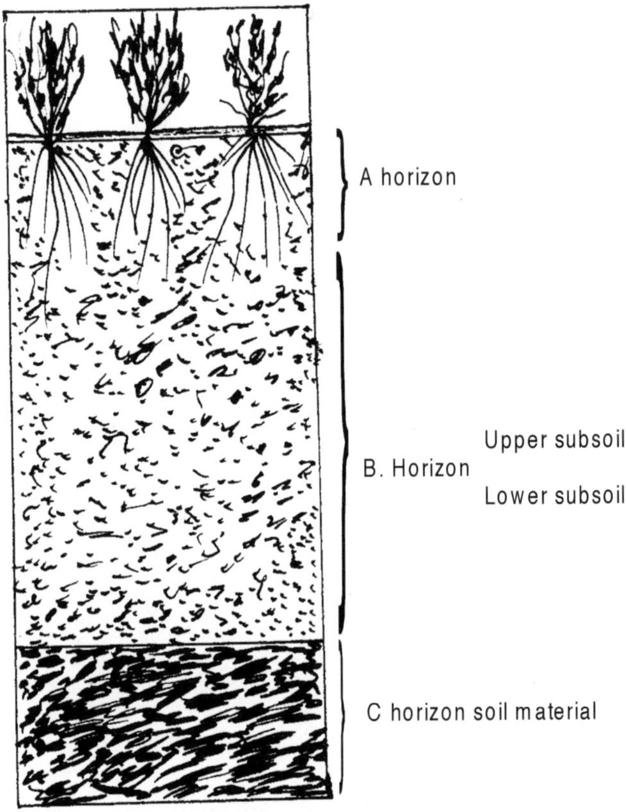

A Representative Soil Profile

relating to the genesis, the profile characters and the normal functioning of this natural body, the soil, will not be satisfactorily solved until we know more about the mineral colloidal nature.

Soil's magnitude of acidity The active acidity of soil solution is ridiculously small even at its maximum.

Soil's magnitude of exchangeable acidity Heavy amounts of limestone are applied for neutralising the acidity. The reason for this heavy application is due to its presence in large part to the magnitude of the exchangeable acidity.

Soil's manifestation of acidity The carbon cycle as it exists and operates in soils, produces a copious amount of carbon dioxide and the development of carbonic acid therefore. Although very weak carbonic acid develops enough active hydrogen to influence markedly the trend of numerous soil transactions.

Soil management Although problems of soil management differ widely with the nature of the soil, the climatic conditions and the types of crops to be grown, yet there are certain fundamental factors that should govern the choice of soil management practices everywhere. Good soil management must be based on the following guiding principles:

1. Selection of the right crops for a given soil or the right soil for a given crop.
2. Maintenance of soil so that it provides suitable conditions for the successful growing of selected crops.
3. Improvement of the productive capacity of soil.
4. Recommendation of only economically profitable husbandry methods.

Soil map To show on maps their distribution in the field.

Soil and its utilization The function of soil survey is to classify locate on a base map, describe the nature of soils as they occur in the field.

Soil materials The unconsolidated residues left behind after much of weathering has taken place, the regoliths are extremely variable from place to place in depth in colour, in texture and in mineral and chemical composition.

Soil material classification Two main groups are recognised (1) Sedentary and (2) Transported.

Soil mechanical analysis In order to throw some light on the physical properties of soil mass at the same time lending itself to an actual separation and percentage evaluation, the analytical procedure followed is mechanical analysis.

Soil micro-organisms antagonistic interrelation Certain bacteria fungi and actinomyces posses qualities of inhibiting the growth of other micro-organisms, especially those that have no concern with soil and its activities.

Soil moisture regulations Soil moisture can be controlled in four situations: (1) runoff (2) percolation (3) weed transpiration (4) evaporation.

Soil moisture It is commonly expressed in terms of pF values, which are logarithmic values for tensions shown in terms of the height, in centimetres, of a water column. The pF numerical values are the same as pH values denoting hydrogen-ion concentrations. The pF value of oven-dry soil is about seven. At the hygroscopic point, it is about 4.5, at the wilting point about 4.2 and at the point of field capacity about 2.5.

Soil movement Heavy or light texture of parent material affects the downward movement of water and thus the development of soil.

Soil nitrogen volatilisation This is most encouraged by poor drainage and lack of aeration. Maintaining the soil in a bare condition and the presence of excessive amounts of mobile nitrogenous compounds are also important factors.

Soil mulch for dry-land farming In the semi-arid regions if much of the precipitation comes in the spring on anyone time, it is practicable to establish the soil mulch near the beginning of the dry season. It will check the surface evaporation of water.

Soil's organic colloids The organisation of the soil humus is in spite of the variable and compels chemical nature is comparatively simple. The same general formula as employed for clays–XCarH (acidoid), can be employed for the colloidal set up of humus appropriately covers the situation.

Soil organisms A vast number of organisms live in the soil. By far the greater number of these belong to plant life and are the forms most important in promoting the final changes in structure and composition which contribute to soil productiveness. Some of these organisms are microscopic and others are as big as rodents.

Soil organisms' activities beneficial to plants In their influence on crop production, soil organisms are of vastly greater help than hindrance to the former. The animal forms convert the insoluble materials to available forms. All these forms have to do something with the process of dissolution! simplification and synthesis etc.

Soil organisms classification They belong to animals and plants. Of these some are macro-organisms and some are micro-organisms belonging to plants and animal groups. Of the animals rodents, insects etc, and among the plants algae, fungi etc. are all very important.

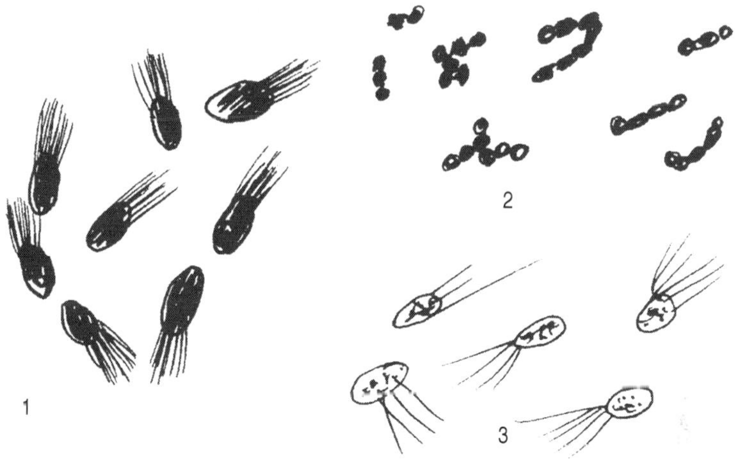

Some Important Soil Organisms for the Nitrogen Cycle in the Soil
1. ***Azotobacter*****; 2. Nitrate bacteria; 3. Nodule bacteria of legumes**

Soil organic matter *See* organic deposits and organic matter.

Soil organic matter and its energy-transfer The micro-organisms of the soil as they grow and multiply, must not only have substance for their tissue synthesis but energy as well. Both of these are obtained in very large degree from the soil organic matter.

Soil organisms and pH of soil Organisms are often affected markedly by fluctuations in the reaction of the soil solution. In most soils, it is ascribed to the factors correlated with pH itself. *See* bacteria.

Soil organic matter sources Main source of organic matter of soil is plant and animal tissues, the former furnishing the initial and major supply. Organic residues, dead grass, litter, iron ores as they, decompose also form part of soil organic matter. Earthworms, centipedes, ants and other forms of animal life- play an important role in the latter translocations. Much of the organic matter also originates from the roots of shrubs, trees and other plants.

Soil-plant inter-relationship The root hairs and rootlets, in most instances are in very intimate contact with the soil colloidal surfaces. By intimate contact, plants may acquire nutrients

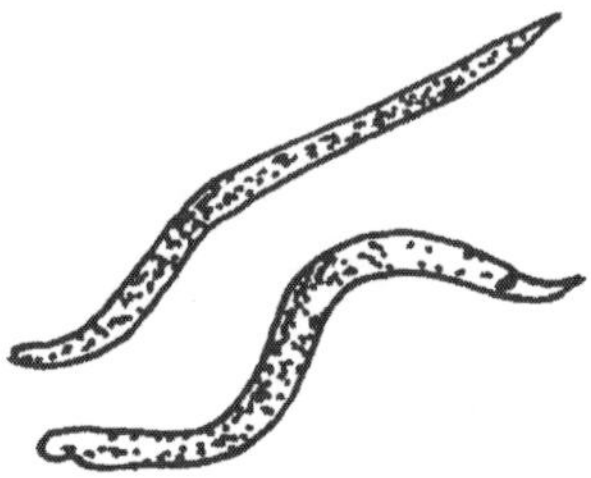

Parasitic nematodes of soil

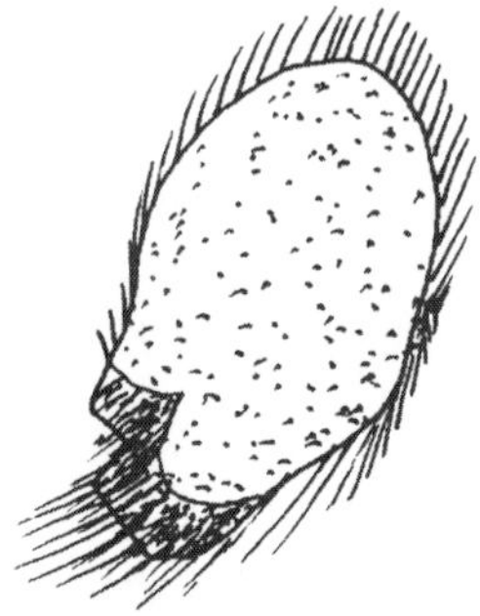

A ciliated protozon of soil

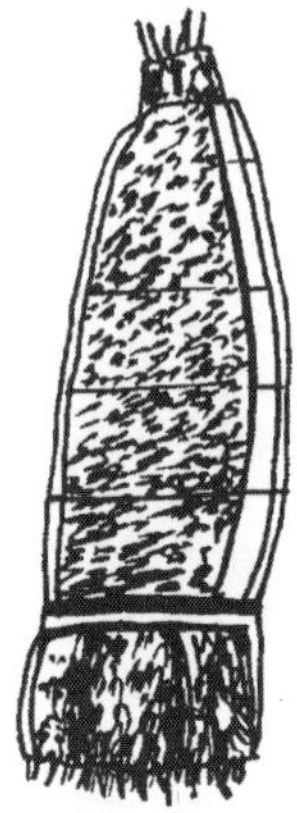

A common rotifer of soil

Common Organisms of Soil

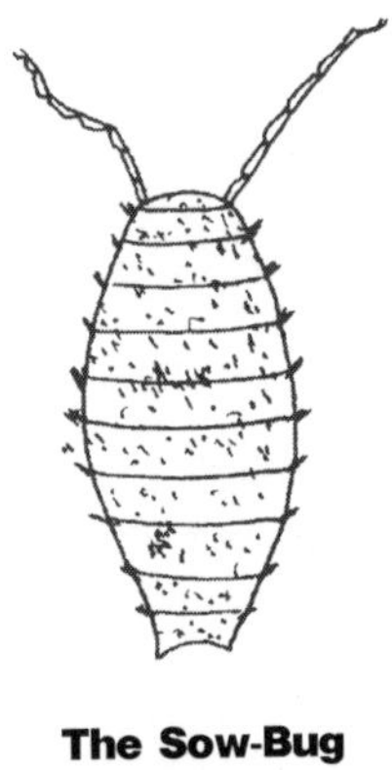

The Sow-Bug
Trachelifus ratfiken

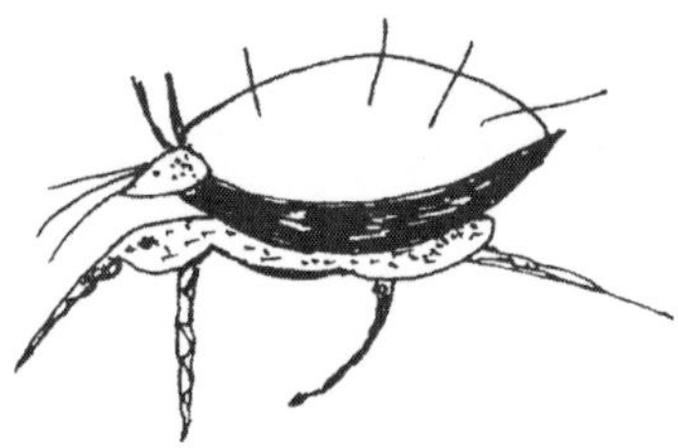

A Mite
Oribata **sp.**

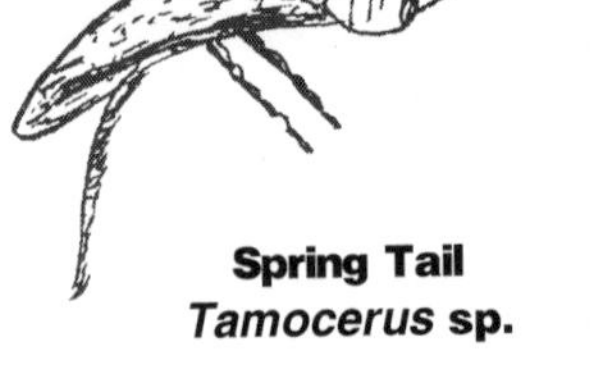

Spring Tail
Tamocerus **sp.**

A melepede
Parajulux **sp.**

The Insects Responsible for the Digestion of Organic Matter of a Soil

before they have chance to enter the soil solution proper and suffer loss by drainage.

Soil Part	*Uncultivated Grazing Land in Per cent*	*Analysis of Sand Pile Collected at a Distance from the Origin in Per cent*	*Analysis of Sand Pile Collection at a Distance of 500 Mile Away from the Origin in Per cent*
Organic matter	1.06	0.33	3.35
Nitrogen	0.06	0.02	0.19
Phosphorous pentaoxide	0.04	Negligible	0.19
Potassium oxide	2.05	1.77	2.58
Sand (all sizes)	79.20	91.80	0.00
Silt and clay	19.60	7.50	9.70
Fine colloids	8.10	5.20	33.40

Soil profile For a study of the soil in its undisturbed condition in the field, a vertical section up to unweathered rock (or up to a depth of about six feet in the case of alluvial soils) is usually examined. This vertical section is called 'soil profile' and it is made up of a succession of horizontal layers (termed 'horizons') which are varying thickness but can be reasonably differentiated on the basis of colour, texture, structure, etc. The profile of a soil can be determined from a specially dug soil pit. They usually show different layers (horizons), from which the soil is classified. *See* also soil horizon. Compare pedon.

Soil pores and adjustment of water When water is applied to a surface of an air dried field soil in good physical condition, it tends to penetrate due to the combined action of gravity and capillarity. When enough water has entered the pores of the soil to satisfy the adsorptive capacity of the surface layer, the excess usually sinks into the lower horizons. Finally it may reach the water table underneath.

Soil reaction One of the important physiological characteristics of the soil solution is its reaction. And since micro-organisms and higher plants respond so markedly to their chemical environment, the importance of soil reaction has long been recognised. Three conditions are possibly of: alkalinity, neutrality and acidity.

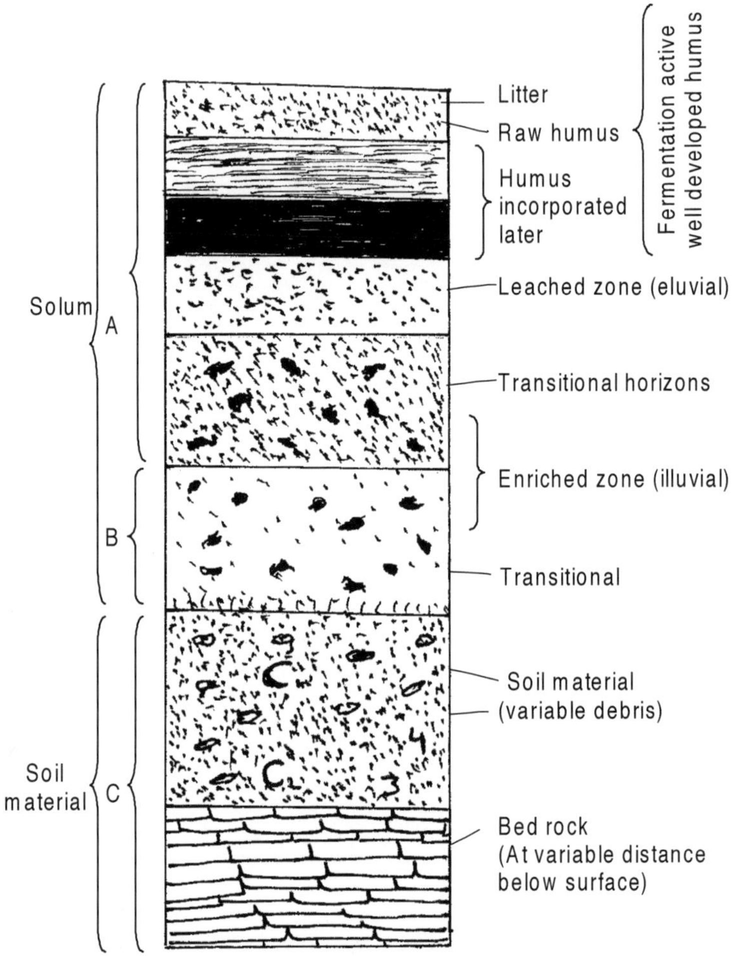

An Ideal Soil Profile Showing All Horizons Distinguised

Soil reaction correlations Certain correlations have been established that are of considerable practical and scientific importance. The interrelations between pH values and (1) exchangeability of calcium and magnesium (2) the solubility of iron, manganese, aluminium and other elements (3) the availability of phosphorus and (4) the activity of soil micro-organisms are the phases that receive the most attention.

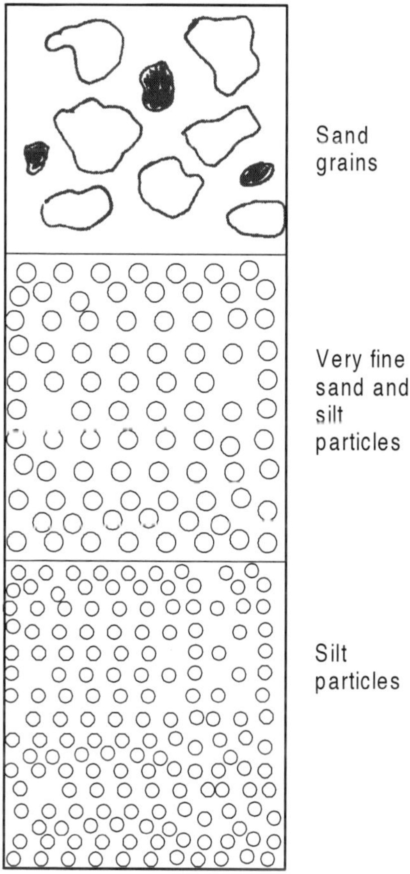

Comparative Sizes of Various Soil Particles

Soil reclamation The excessive accumulation of salts is injurious to crop growth, hence the need for removal of alkali. The reclamation techniques for different categories of saline and alkali soils differ widely, but they conveniently be grouped as follows:

1. Chemical methods, of reclamation such as use of calcareous amendments.
2. Mechanical methods, such as lowering of water table, drainage, leaching, mechanical shattering of indurated clay pan and scrapping.

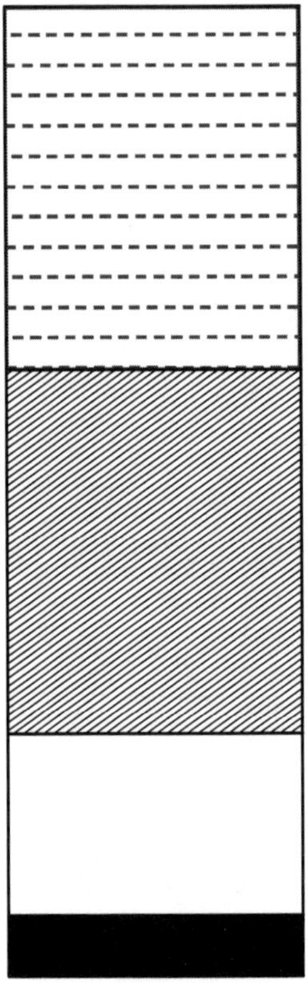

Volume Composition of a Silty Loam Surface Soil

3. Growing of salt or alkali-tolerant crops for reclamation. The reclamation of saline solid requires a permanent lowering of water table to five or six feet below the surface. The provision of a network of deep cuts at right angles to the slope of the land serves the purpose. This should be followed by leaching with good quality irrigation water. Usually two or three leachings suffice to remove all harmful salts. Leaching also drives down considerable quantities of salts below the ordinary root zone. The growing of salt-tolerant crops, such as rice, barley, sugarcane and castor helps as gradual reduction of salt concentration.

Soil's retention of water When water is adsorbed by a soil or by any other porous medians, energy as heat is liberated. This is easily demonstration by moistening a dry soil under controlled condition.

Soil-rollers On light land, a set of rolls weighing up to 12 qwt. and 10 to 12 feet wide is required to provide full load for a small tractor. The roller needed for heavy land should weigh 17 qwt. or more. The 'Cambridge' roller with fluted surface is very effective as a cold crushing unit.

Soil scoop It consists of a shallow box-like construction with an open front. On this open front, a sharp blade is fixed and two hooks are attached, to which the bullocks are yoked. Two wooden handles are provided at the back. The capacity of the soil scoop is 2 to 3 c.ft. When the soil is to be carried to fairly long distance, this implementation is used.

Soil separates *See* mechanical analysis.

Soil series A collection of soils that have very similar profiles and are developed over the same parent material. Soil series are subdivisions of soil groups determined by local geological conditions. Members of soil series differ only in texture of there A horizon (surface horizon). The soil series are used as a basis for soil mapping. They are differentiated by being given local names. Compare soil type.

Soil solution The liquid content of the soil, soil-solution, consists of water containing varying quantities of dissolved mineral matter, carbon dioxide and oxygen. It is the soil solution through which mineral nutrients, nitrogen, water and some carbon dioxide enter the plant. The soil solution exists in a state of minute subdivision and dispersion. Moreover, it is differentially absorbed by the colloidal surfaces, part being held tightly and having little movement, while some suffers but slight restraint and circulates more or less freely. The soil solution is also changeable, varying as to the gross amount present in any soil, as to total concentration of dissolved salts, as well as to the amount and proportion of specific constituents.

Soil's specific heat The specific heat of a substance may be defined as its thermal capacity compared with that of water. It is expressed as ratio of the quantity of heat required to raise the temperature of a given amount of certain substance from 15 to 16°C.

Soil solution extracts Soil solution extracted with various solvents, such as alcohol, oil or water, are probably not identical in composition and concentration to the soil solution in contact with root hairs because the plant by excreting carbon dioxide and other substances, helps in dissolving plant nutrients.

Soil sterilants These chemicals make the soil sterile (incapacitate the soil to support vegetation) for varying lengths of time depending on the chemical used, the amount applied, rainfall, soil type, organic matter content, etc. The inorganic chemicals borates and chlorates, the triazines, simazine and atrazine, the urea herbnicides such as diuton, fenuron, monuron and neburon are some soil sterilants.

Soif structure The arrangement of particles in the soil mass is known as soil structure. In dry, coarse sand, the structure is called separate grain structure. In most of the other soils, the primary soil particles are found in a state of aggregation. These compound aggregates vary in size and shape depending on the nature of the clay minerals contained in them and the processes of cultivation and soil formation to which they have been subjected. The term structure is used to refer to the arrangement as well as size of the particles which must exert considerable influence. The term structure is used to refer to the arrangement or grouping of the soil materials into aggregates.

Soil structure types In sand, the soil masses are very large, irregular and featureless as regard to the characteristic aggregation. But within these two there are four structural types commonly recognised, which are platy, prismatic, blocky and granular.

Soil sulphites Many organic compounds, especially those of a nitrogenous nature, carry sulphur, which appear in the simple forms as decay progresses.

Soil survey For a detailed planning of proper land use, it is necessary to chart soil resources by a systematic survey. In fact, soil survey is the starting point for all soil research and long term planned land use.

Soil survey and its utilisation The function of a soil survey is to classify, locate on a base map and describe the nature of soils as they occur in the field.

Soil temperature The temperatures needed for seed germination and root growth vary with crops and varieties. It is well known that wheat, barley and peas, which are grown as winter crops in India, can germinate at a relatively lower temperature than maize, cotton, groundnut and melons. In formation of soils, it has a major role as it influences the process by differential expansion of miners–frost action and exfoliation.

Soil temperature and atmospheric radiations A transfer of energy through space by means of waves or pulsations, which are transformed into heat values, when they encounter an obstructing body.

Soil temperature by conduction The surface layers of the soil, where energy is being continually received, are subject to sudden and often drastic fluctuations in temperature. This is due in part of variations in solar isolations and in part to losses to the atmosphere.

Soil temperature of bare and cropped soil Whether the soil is bare or covered with vegetation is another factor that markedly influences the amount of insulation received. The effect of forests is universally recognised while even an ordinary field crop, such as some grasses have a very noticeable effect, especially upon temperature fluctuations.

Soil temperature fluctuations The surface layers of a soil, where energy is continually being received, are subject to sudden and often drastic fluctuations in temperatures. This is due in part to solar variations of insulations and in part to the losses to the atmosphere. The latter is due to (1) heat conduction (2) to the atmosphere (3) radiation into space and (4) to the cooling effects of moisture evaporation of the soil.

Soil temperature insulation contingencies The amount of solar insulation accorded to any particular locality depends fundamentally upon climate. But the amount of heat entering the soil is in addition, depends largely on (1) the colour (2) the slope (3) the vegetative covers of the site under consideration. Dark soils absorb more heat energy than light coloured ones, and that red and yellow soils will show a more rapid temperature rise than those that are white. It has also been shown that near the angle of incidence of the Sun's rays approaches the perpendicular, the greater will be the absorption.

Soil temperature variation due to conduction A molecular transfer of energy from one part of a particular body to another. The energy is conceived as communicated progressively from molecule to molecule.

Soil test A soil test is a scientific assessment by chemical or other analytical methods of the quantities of plant nutrients present in a particular soil or of other soil conditions affecting crop growth and yield. Soil sampling must be done properly.

Soil testing kits There are the handy equipments by use of which soil tests can be done quickly and conveniently. There are very cheap and every farmer can have a set of them for his use.

Soil testing laboratory By making soil tests, we can know the available quantity of the nutrients in the soil and the quantity of a particular nutrient required when growing a particular crop. The progressive formers send their soil samples to these laboratories for analysis. The laboratory in turn analyse the soil samples and report the quantity of available nutrient elements and also give out the plan for taking crops with the proposed scheme of fertilization of the field.

Soil testing operation Under the terms of Operational Agreement No. 4 between the Government of India and the U.S. Technical Cooperation Mission, 16 Soil Testing Service Laboratories were set up in 1956 in sixteen States then existing which were subsequently increased to 24 under Operational Agreement Nos. 1 and 2.

Soil test reports Besides indicating the quantities of principal nutrients contained in the soil, the reports also contain specific recommendations as to organic matter and fertilizer applications.

Soil texture The proportion of the various particle sizes in a soil.

Soil texture and capillary water As a result of its energy relation the capillary water is the only fluid water bearing solutions, that remains in the soil for any length of time if drainage is satisfactory.

Soils' textural names Four fundamental textural groups of soils are recognised: gravels, sands, loams and clays. On the basis of these other class names are derived.

Soil texture and organic matter The texture of soil, other factors being constant, seems to influence the percentage of humus and nitrogen present. A sandy soil usually carries less organic matter and nitrogen than those of a heavier nature.

Soil texture and percolation The finer and texture of the soil, the slower will be the rate of gravitational water flow. In sandy soils, pores are large, the amount of colloidal matter low and conduction easy. In heavy soils, the pore spaces are small and clogged with colloidal matter and the constrictions many, hence percolation slow.

Soil texture in relation to plant growth The particle groups (*i.e.* separates) differ not only a particle size but also in their bearing on some of the principal factors affecting plant growth, such as (*i*) movement and availability of water, (*ii*) soil aeration, (*iii*) workability, and (*iv*) content of plant nutrients.

Soil water Water forms part of plant food, being used directly as such or after being split up into its component elements. Nearly 90 per cent of plant tissues consist of water. It serves as a solvent and carrier of plant nutrients. It keeps plant cells turgid and regulates their temperature.

Soil water's biological classification From the point of view of energy values of soil water and its movement in the soil, it is evident that not all moisture present is available or suitable for the rapid vegetative growth of a plant. Further tentative divisions of soil water may be made, as unavailable, desirably available and superfluous.

Soil water calculation This is done on percentage of water present in a sample of soil. A sample is thoroughly mixed, is weighed into a suitable dish and air dried. The sample is then cooled in a desiccator and weighed. The loss in weight is weight of water. The moisture is then calculated as a percentage based on the dry weight of the soil.

Soil water evaporation and temperature Evaporation of soil water is caused by an increased molecular activity and requires the expenditure of certain amount of energy, which results in a cooling effect especially at the surface where most of the evaporation occurs. It requires 265 kilograms calories to evaporate 1 pound of water at 68°F. This is sufficient to lower the temperature of a cubic foot of a mineral soil at optimum moisture of about 28°F.

Soil water forms Three kinds of water are found and are related to soil in an intimate way (1) hygroscopic (2) capillary and (3) gravitational. The classification in universal.

Soil water movement and availability Sandy soils are very permeable and well drained but less water retentive and, on that score, they need more frequent irrigations than the clayey soils. The clay particles make the soils more retentive of water. But they hold water so tenaciously that at the same moisture level, a plant wilts earlier in clay soils than in the sandy types.

Soils with unchanged parent material Such soils do not possess any well-developed characteristic produced by any soil-forming force. These soils include tundra soils, podzols, praries soils and desert soils. These are also called azonal soils.

Soil's weight of living matter The total weight of living matter including plant roots, in an acre furrow slice of a representative mineral soils may be about 5000–10,000 pounds.

Soil's weight per cubic foot When the volume weight of a soil is known, its dry weight in pounds to the cubic foot may be found by multiplying by 62.42, the weight one cubic foot of water.

Soil type A subdivision of a soil series based on the texture of A horizon. Soil types are the most specific classification used in description and mapping.

Sol A colloid consisting of solid particles distributed in a liquid medium. A wide variety of sols are known; the colours often depend markedly on the particle size. The term aerosol is used for solid or liquid phases dispersed in a gaseous medium. When a small amount of colloidal clay is dispersed in a much larger quantity of water, a condition designated in colloidal chemistry is as a sol exists. When this relationship is reversed, a relatively small quantity of water is present, a viscous condition results.

Solan Stam F-34 and sulphuric acid.

Solonchak (white alkali soil) An intrazonal saline soil formed where hot arid conditions cause evaporation of soil moisture, bringing dissolved salts to the surface by capillarity from the saline ground water. This results in a greyish crusty appearance of the soil profile.

Solonetz (black alkali soil) An intrazonal saline soil formed where relatively heavy or periodic rainfall leaches salts out of the surface layer into the lower alluvial B horizon. The surface horizons become structureless with dispersed humus and clay, which gives the soil a dark appearance. Compare solonchak.

Solubility of phosphates and carbon dioxide If we take the tricalcium phosphate as in insoluble soil phosphate, a reaction may be shown as:

$$Ca_3(PO_4) + 4H_2O + 4CO_2 = CaH_4\ (PO_4) + 2Ca\ (HCO_3)_2$$

Solubility of potash by carbonic acid The colloidal and easily replace combinations of the calcium are indicated by a generalised formula. that symbolises the absorbed condition in which much of it exists in soils:

$$2KAlSiO_8 + CO_2 + 2H_2O = H_4Al_2Si_2O_9 + K_2CO_3 + 4SiO_2$$

Solution A form of chemical weathering in which mineral materials are dissolved by water, weak acids in the water, such as carbonic acid from rainfall continuing carbon dioxide from the air or humic acid from decayed organic material, increase the amount of solution, as do greater amounts of water flowing or seeping past the minerals.

Solvent A liquid capable of dissolving other materials (solids, liquids or gases) to form a solution. The solvent is generally the major component of solution.

Solvolysis A reaction between a compound and the solvent in which it is dissolved. *See* also hydrolysis.

Specific gravity of soils It is generally determined by means of picnometer, a bottle fitted with a perforated ground-glass stopper and accurately calibrated. Calculation is done as:

1.	Weight of picnometer	22.275 gms
2.	Volume of picnometer	50.00 c.c.
3.	Weight of picnometer +5 gms of absolutely dry soil X gms of water	73.525
4.	Weight of X gms of water	28.255
5.	Weight of picnometer + 5 gms of absolutely dry soil water displaced	45.120
6.	Specific gravity	5.0
		4.07–1.22

Speed It is the distance covered by a moving body in unit time.

Spherometer This scale is so called as it is used for measuring the radius of curvatures of spherical surfaces (convex or concave).

Sprinkler irrigation The watering of crops by using plastic or metal pipes to which fine nozzles are attached. The system is very flexible and takes a number of forms, *e.g.* water under pressure is sprayed from a rotating jet, or pipes fixed above the crops create artificial showers when required.

Stages of soil development The successive stages in soil development are called infancy, youth, maturity and old age. For attaining maturity and old age, positional stability of soil extending over a long period is essential.

Stainless steel *See* steel.

Standard cell A voltaic cell whose e.m.f. is used as a standard. *See* Clark cell, Weston cadmium cell.

Standard electrode A half cell used for measuring electrode potentials. The hydrogen electrode is the basic standard but, in practice, colomel electrodes are usually used.

Standard solution A solution that contains a known weight of the reagent in a definite volume of solution. A standard flask or volumetric flask is used for this purpose. The solutions may be prepared by direct weighting for primary standards.

Standard temperature An internationally agreed value for which many measurements are quoted. It is the melting temperature of water 0°C (263.15k).

Stannane (tin (IV) hydride, $SnHO_4$) A colourless poisonous gas prepared by the action of lithium aluminium hydride on tin (II) chloride. It is unstable, decomposing immediately at 150°C. Stannane acts as a reducing agent.

Stannous compounds Compounds of tin (II).

Starch A polysaccharide that occurs exclusively in plants. Starches are extracted commercially from maize, wheat, barley, rice, potatoes and sorghum. They exist in the plant cells as granules dispersed in the cytoplasm. The starches are storage reservoirs for plants; they can be broken down by enzymes to simple energy needs. Starch is a dietary component of animals. In man, it is digested by salivary and pancreatic amylase then further degraded by maltase to yield glucose, which may be stored as glycogen (animal starch). Excess starch, *i.e.* above the maximum liver and muscle storage capacity, is converted to lipids and stored as fat. Starch is not a single molecule but a mixture of amylase (water-soluble, blue colour with iodine) and amylopectin (not water-soluble, violet colour with iodine). The composition is amylase 10–20 per cent, amylopectin 80–90.

States of matter The three physical conditions in which substances occur; solid, liquid and gas. The addition or removal of energy (usually in the form of heat) enables one state to be converted into another. The major distinctions between the states of matter depends on the kinetic energy and are very loosely packed; kinetic energy and separation of particles in liquids are intermediate. Solids have fixed shapes and volumes, *i.e.* they do not flow, like liquids and gases, and they are difficult to compress.

Steam distillation A method of isolating or purifying substances by exploiting Dalton's law of partial pressures to lower the boiling point of the mixture. When two immiscible liquids are distilled, the boiling point will be lower than of the more volatile component of water. The method is particularly useful for recovering materials from arry mixture.

Stratification (Biogeography) The layered arrangement of vegetation within a plant community, *e.g.* mature woodlands may show the following layers:

1. Scattered emergent trees.
2. General canopy.
3. Under-storey of shrubs and sapling.
4. Herb or field layer.
5. Ground flora *e.g.* mosses.

Strength of adsorption The replacing power of the various cations is H > Ca > Mg > K > names as action is also a very important factor. This order is related to the valency of the ions and their degree of hydration.

Structure The arrangement and positioning of the rocks of the earth's crust, including the hardness and texture of the rocks and their form before erosion.

Structural formula The formula of a compound showing the numbers and types of atoms present, together with the way in which these are arranged in the molecule.

Structural isomerism *See* isomerism.

Structural management This sort of management of a silicate clayey soil is not such a simple problem as that of sandy one. In the latter, the plasticity and cohesion are never great even after the addition of large amounts of organic materials. In clays and similar soils of temperate regions, however, the potential plasticity and cohesion are always high due to the presence of large amounts of complex hydrated aluminium silicates in a colloidal solution.

Structure of rocks Arrangement and positioning of the rocks of the earth's crust including the hardness and texture of the rocks and their form before erosion.

Structure of soil It refers to the arrangements or grouping of the soil material into aggregates. Apparently it affects the soil condition such as aeration, water movement, and heat transfer and similar other properties, which will be much, affected by structure of soil.

Sublimation The change of stage of a substance directly from solid to gas (or vice versa) without an intermediate liquid stage. In meteorology, it is most important for the formation of ice crystals directly from water vapour within the atmosphere, without an intervening liquid state, when the air temperature is below freezing.

Subsoil A layer at the base of soil, which underlies the topsoil. In general, it corresponds with the B horizon in the soil profile. The term is commonly used in agriculture to describe the mineral not disturbed by normal ploughing. Subsoil is less exposed to soil forming processes. As a result, it has a higher mineral content. It also has correspondingly less organic matter (humus), fewer soil organisms, and less penetration by plant roots. Compare topsoil. *See* also soil.

Subsoil placement By this method, manures are placed at depth by deep plough. As usual in case of acidic soils, they become poor in the available nutritive elements and for this reason plants growing in such soils do not exhibit a good growth. To meet this discrepancy with advantage, fertilizers are placed in the subsoil region.

Sucrose (cane sugar) A sugar that occurs in many plants. It is extracted commercially from sugar cane and sugar beet. Sucrose is a disaccharide form from a glucose unit and a fructose unit. It is hydrolysed to a mixture of fructose and glucose by the enzyme invertase. Since this mixture has a different optical rotation (laevorotatory) than the original sucrose, the mixture is called invert sugar.

Sugar (saccharide) One of a class of sweet-tasting simple carbohydrates. Sugars have molecules consisting of a chain of carbon atoms with –OH groups attached, and either an aldehyde or ketone group. Monosaccharides are simple sugars that cannot be hydrolysed to sugars with fewer carbon atoms. Two or more monosaccharide units can be linked in disaccharides.

Sugar acids An acid formed from a monosaccharide by oxidation. Oxidation of the aldehyde group (CHO) of the aldose monosaccharides to a carboxyl group (COOH) gives an aldonic acid; oxidation of the primary alcohol group (CH_2OH) to COOH yields uronic acid; oxidation of both the primary alcohol and carboxyl groups gives an aldaric acid. The uronic acids are biologically important, being components of many polysaccharides, for example, clucuronic acid (from glucose) is a major component of gums an cell walls, while galacturonic acid (from galactose) makes uppectin. Ascorbic acid or vitamin C is an important sugar acid found universally in plant tissues, particularly in citrus fruits.

Sugar alcohol An alcohol derived from a monosaccharide by reduction of its carbonyl group (CO) so that each carbon atom of the sugar has an alcohol group (OH), for example, glucose yields sorbitol, common in fruits, and man nose yields mannitol.

Sulpha drugs *See* sulphonamide.

Sulphate A salt or ester of sulphuric (VI) acid.

Sulphate of ammonia Development of acidity by this fertilizer is not serious. A little increase in the amount of lime added or an alteration of nitrate or soda with the ammonium sulphate will counteract this tendency.

Sulphite A salt or ester of sulphurous acid.

Sulphonamide A type of organic compound with the general formula $R.SO_2\ NH_2$. Sulphonamides, which are amides of sulphonic acids, are active against bacteria, and some are used in pharmaceuticals ('sulpha drugs').

Sulphonation A reaction introducing the $-SO_2OH$ (sulphonic acid) group into an organic compound. Sulphonation of a aromatic compounds is usually accomplished by refluxing with concentrated sulphuric acid for several hours. The attacking species is SO_3 (sulphur (VI) oxide) and the reaction is an example of electrophilic substitution.

Sulphonic acid A type of organic compound containing the $-SO_2OH$ group. The simplest example is benzenesulphonic acid ($C_6H_5SO_2OH$). Sulphonic acids are strong acids. Electrophilic substitution can introduce other groups onto the benzene ring; the $-SO_2OH$ group directs substitutes into the 3-position.

Sulphur A low melting non-metallic solid, yellow coloured in its common forms; the second member of groups VI of the periodic table. It has the electronic configuration $(Ne)3S^2 3P^4$. Sulphur occurs in the elemental form in Sicily and some southern states of the USA and in large quantities in combined forms such as sulphide ores (FeS_2) and sulphate rocks ($CaSO_4$). It forms about 0.5 per cent of the earth's crust. Elemental sulphur is extracted commercially by the Frasch process in which superheated steam is forced down the outer of three concentric tubes leading into the deposit. It is found in the plant tissue in quantities of about 0.05 to 1.5 per cent. Plants stake it in the form of SO_3^- and SO_4^-. Except on the tips of leaves, the whole leaves suffer from chlorosis. Growth of plant gets much retarded. The new leaves start becoming light green. Sulphur forms an important constituent of straw and plant stalks. Onion, garlic, cabbage, radish, turnip and some legumes like groundnut, gram and alfalfa contain a high percentage of sulphur deficiency, the new leaves turn yellow, roots and stems become abnormally long and develop woodiness. In fruit trees, the fruits become light green mis-shaped, thick skinned and less juicy. Removal of sulphur from soil by drainage is much more as compared to phosphate, which is negligibly drained out of soil. However, considerable sulphur in the sulphate form is added to the soil every year in rain water.

Sulphuric (VI) acid (H_2SO_4) A colourless only liquid manufactured by the contact process. The pure acid is diluted by adding it slowly to water, with careful stirring. Concentrated sulphuric acids acts as an oxidizing agent, giving sulphur (IV) oxide as the main product and also as a dehydrating agent. The diluted acid acts as a strong dibasic acid, neutralizing bases and reacting with active metals and carbonates to form sulphates. Sulphuric acid is used in the laboratory to dry gases (except ammonia) to prepare nitric acid and ethane and to absorb alkalenes. In industry, it is used to manufacture fertilizers (*e.g.* ammonium sulphate), rayon and detergent to clean metals, and in car batteries. *See* contact process.

Sulphurous acid (sulphuric (IV) acid, H_2SO_3) A weak acid found only in solution made by passing sulphur (IV) oxide into water. The solution is unstable and smells of sulphur (IV) oxide. It is a reducing agent, converting iron (III) ions to iron (II) ions, chloride ions and orange dichromate (VI) ions to green chromium (III) ions.

Super over, above.

Supercooling The process whereby water can be cooled well below its freezing point (0°C) and yet remain in the liquid state. This is common in the atmosphere as a result of the rarity of freezing nuclei and is important in producing certain weather phenomena.

Superficial deposit A deposit of material laid down above the solid bedrock as a result of having been transported to its present position and deposited by geomorphological processes. Superficial deposits include glacial and fluvioglacial drift, alluvium (*e.g.* river terraces), estuarine and deltaic deposits, and windblown sand and loess.

Superheating The raising of a liquid's temperature above its boiling temperature by increasing the pressure.

Superimposed drainage (superimposition) A river network that runs illogically across resistant and non-resistant rocks alike. *See* also discordant drainage.

Supersaturation The condition where the air is cooled to below its dew point without condensation taking place, *i.e.* the air contains more water vapour than is required to saturate it.

Supplementary units The dimensionless units–the radian and the steradian used along with base units to form derived units. *See* also SI units.

Surface creep Grains larger than about 0.5 mm in diameter but smaller than 1.0 mm are too heavy to be moved in saltation but are pushed or speed along the surface by the impact of particles in saltation to form surface creep. About 90 per cent of the total soil movement in wind erosion is below the height of 12 inches and about 50 per cent of it is within two inches of the ground level. The control of wind erosion is mainly based on reduction or elimination of movement in saltation.

Surface drains Those drains which are constructed for removal of excessive surface or irrigation water.

Surface irrigation There are four variations under this method, *viz.* (*i*) flooding, (*ii*) bed or border method, (*iii*) basin method, and (*iv*) furrow method.

Surface tension The attractive force of water molecules for each other is manifest at all liquid air interfaces by a phenomenon called surface tension. This is well understood stress, which is evidenced by the tendency of the film to contract, reduces the water-air interface in any particular case to the least possible area and in so doing develops a negative pressure. Moreover, it is well known that this pressure produced by surface tension, varies directly with the curvature of the film.

Surface water Surface water is provided by the flowing waters of rivers or from the still waters of tanks, ponds, lakes or artificial reservoirs. Irrigation from rivers is mainly through canals draws from dams constructed across the rivers and is mainly seasonal, *i.e.* during the flood season.

Surfactant A substance that lowers surface tension, *e.g.* a soap or other detergent.

Suspension A method of transport of solid particles in which they are carried in a fluid, such as moving air or water. *See* also suspension load.

Suspension load (suspended load) That part of the material transported by moving water, such as a river, that is carried by the fluid, the size of particles that can be carried in suspension increases with the velocity of flow and its turbulence and tractive force. Some particles may only remain in suspension for short distances. Generally suspension load accounts for 70–90 per cent of a river's total load of solids.

Syncline A basin-shaped structure in sedimentary rocks in which the strata dip towards each other. Synclines alternate with anticlines and result from compressional forces acting on the rocks. In strongly folded rocks, the limbs of the syncline are steep but some synclines form broad shallow basins.

T

Tannin One of a mixed group of substances which, as defined by industry, combine with hide to form leather. Tannins are also used in dyeing and ink manufacture. Many plants accumulate tannins particularly in leaves, fruits, seed coats, bark, and heartwood.

Tannins precipitate proteins and hence inactivate enzymes; they are therefore segregated in cell vacuoles, organelles, or cell walls. Chemically they are polymers derived either from carbohydrates and phenolic acids by condensation, reactions, or from flavonoids.

Tartaric acid A crystalline hydroxycarboxylic acid with the formula:

$$HOOC\,(CHOH)_2\,COOH$$

Its systematic name is 2,3-dihydroxybutanedioic acid. *See* optical activity.

Tartrate A salt or ester of tartaric acid.

Tautomerism Isomerism in which each isomer can convert into the other, so that the two isomers are in equilibrium. The isomers are called tautomers. Tautomerism often results from the migration of hydrogen atoms. *See* ketoenol tautomerism.

Tectonic Denoting the forces responsible for the deformation of the earth's crust, including the growth of mountain ranges and volcanism, and the landforms, structures etc. resulting from these forces, *e.g.* tectonic earthquakes. *See* also plate tectonics.

Temperature An index of the degree of heat of a body substance or medium such as the atmosphere. It is a fundamental component meteorology and climatology and every weather station takes various temperatures reading every day. It is influenced by many factors as whatever effect the weather.

Temporary hardness A type of water hardness caused by the presence of dissolved calcium, iron and magnesium hydrogen carbonates. This form of hardness can be removed by boiling the water. Temporary hardness arises because rainwater combines with carbondioxide from the atmosphere to form weak carbonic acid. This reacts with carbonate rocks, such as limestone acid, to produce calcium hydrogencarbonate which then goes into solution.

Terpene One of a complex group of lipids based on the hydrocarbon skeleton C_5H_8 (isoprene). Monoterpenes are built from two C_5 residues ($C_{10}H_{16}$), diterpenes from four etc. The C_{10} to C_{20} terpenses are present in essential oils, giving the characteristic scent of some plants (*e.g.* mint). Some terpenoid substances are physiologically active, *e.g.* vitamin A.

Terra rossa A red intrazonal soil that occurs on limestones in Mediterranean climates. It generally has a clay-loam texture. Terra rossa soils are the insoluble residues from the chemical breakdown of limestone. Where these residues accumulate, in crevices and depressions, the dissociation of clays produces ferric oxides, which are responsible for the red coloration, and leaching causes acidity. The development of terra rossa thus requires a wet season of moderate rainfall.

Tetrahedral compound A compound such as methane, in which an atom has four bonds directed towards the corners of a regular tetrahedron. The angles between the bonds in such a compound are about 109°.

Tetrachloromethane carbon tetrachloride Note that 'tetra-chloromethane' is the systematic name, and is preferred in organic chemistry.

Textural classes The varying proportions of particles of different size groups in a soil constitute what is technically termed as its textural class. Soils in the field are generally made up of varying mixtures of sand, silt and clay and are recognised as a number of intermediate soil-texture classes. The principal classes in the order of increasing content of finer particles (*i.e.* silt and clay) are sand, loamy sand, sandy loam, loam, silt loam, silt, clay loam and clay.

Tillage Land used for growing crops has to go under certain operations on the soil, such as ploughing, harrowing, hoeing etc. This is a part of the preparation of land for sowing and cropping. *See* crop science.

Tillage and granulation Tillage operations be carefully timed in a heavy soil so that granulation may be encouraged to the fullest degree. Here the organic matter is the main concern.

Tillage and soils' volume weight Differences in volume weight of surface soils may be accentuated by methods of farm management. The addition of farm manure in large amounts tends to lower the weight as does also a grass sod.

Tin A silvery white metal. It is rarely found in its native state and is derived chiefly from cassiterite (tin oxide, SnO_2).

Tin (II) oxide (SnO) A dark green or black solid. It can also be obtained in an unstable red form, which turns black on exposure to air. Tin (II) oxide can be made by precipitating the hydrated oxide from a solution containing tin (II) ion and dehydrating the product a+100°C.

Titanium A transition metal that occurs in various ores as titanium (IV) oxide and also in combination with iron and oxygen. It is extracted by conversion of titanium (IV) oxide to the chloride, which is reduced to the metal by heating with sodium. Titanium is reactive at high temperatures. It is used in the aerospace industry as it is strong, resistance to corrosion, and has a low density. It forms compounds with oxidation states +4, +3, and +2, the +4 state being the most stable.

Titration *See* volumetric analysis.

Tollern's reagent A solution of the complex ion $Ag(NH_3)_2^+$ produced by precipitation of silver oxide from silver nitrate with a few drops of sodium hydroxide solution, and subsequent dissolution of the silver oxide in aqueous ammonia. Tollen's reagent is used in the 'silvermirror test' for aldehydes.

Topsoil The well-developed part of the soil at or near the ground surface (*i.e.* the A horizon). Topsoil is the vital zone for plant growth. The term is generally used for the one above the subsoil that is moved during cultivation. Compare subsoil.

Transfer of soil material It involves the transfer of disintegrated parent material from one place to another. The formers are called sedentary and the latter transported.

Transported *See* transfer of soil material.

Transported soil material These are recognised on the basis of according to the agencies of transportation and deposition through water, air and winds.

Triple point The only point at which the gas, solid and liquid phases of a substance can coexist in equilibrium. The triple point of water (273.16k at 101 325 pa) is used to define the kelvin.

Tritiated compound A compound in which one or more 1 H atoms have been replaced by tritium (3H) atoms.

Tritium Symbol: T A radioactive isotope of hydrogen of mass number 2. The nucleus contains 1 proton and 2 neutrons. Tritium decays with emission of low-energy beta radiation to give ^3He. The half-life is 12.3 years. It is useful as a tracer in studies of chemical reactions. Compounds in which ^3H atoms replace, the usual ^1H atoms are said to be tritiated.

Trivalent (tervalent) Having a valency of three.

Tropical climate The group of varying climatic types found within the tropics. Each climatic classification has variations but it is generally agreed that there are five basic types all having one thing in common that it has no winter.

Tuff A deposit formed from consolidated volcanic ash, usually having a grain size of less than 4 mm. Some tuffs have been laid down under water as they are fond inter-layered with ordinary sedimentary rocks. There are three main forms of tuff–(1) Vitric tuff is composed chiefly of volcanic glass and results from an explosive eruption of liquid lava. (2) Crystal tuff is composed of detached crystals and results from lava that has nearly crystallized at the time that it is thrown out in an eruption. (3) Lithic tuff is composed chiefly of angular fragments produced by the shattering of solid rock during a volcanic explosion.

Tundra soils These are found below the belt of perpetual snow and ice at very high altitude in the Himalayas and other mountain ranges.

Tungsten A metal obtained from veins of the minerals–scheelite and wolframite. The metal has a high melting point and is very strong and hard.

Turbidity 1. The opacity of the atmosphere that results from the presence of particles such as dust and smoke. 2. The muddiness of water resulting from suspended sediment.

Types of saline and alkali soils Three broad types are recognised: saline soils, saline alkali soils and alkali soils.

Type of soil Denotes its structure, texture, organic matter content, its infiltration capacity and permeability greatly affects soil loss and runoff. Heavy or alkaline soils are highly susceptible to erosion. They get easily puddled and very quickly form an impermeable infiltration.

U

Unit The unit of a physical quantity is the convenient and definite magnitude of that quantity with which we compare other magnitudes of that quantity. Each physical quantity requires a special unit which are of two kinds (1) fundamental units and (2) derived units.

Unit cell The smallest group of atoms, ions or molecules that, when repeated at regular intervals in three dimensions, will produce the lattice of crystal system. There are seven basic types of unit cells, which result in the even crystal systems.

Unit of classification To classify soils into units of a uniform system of classification with uniform nomenclature in order to correlate them with those in other areas.

Units of force:

	Systems	*Absolute Units*	*Gravitational Units*
1.	C.G.S.	1 dyne^{-1} gram × 1 cm/sec^2	1 gm. wt. G dynes = 981 dynes
2.	M.K.S.	1 newton = 1 Kg × 1 metre/sec^2	1 kg wt. G newtons = 9.81 newtons
3.	F.P.S.	1 poundal = 1 pound × 1 ft/sec^2	1 lb wt. G poundals = 32 poundals

Units of power

	Systems	Absolute Units	Gravitational Units
1.	C.G.S.	Erg/sec	gm-cm/sec
2.	M.K.S.	Joule per sec/watt	Kg-metre/sec
3.	F.P.S.	ft poundal/sec	ft lb/sec

Unit processes (chemical conversions) The recognized steps used in chemical processes, *e.g.* alkylation, distillation, hydrogenation, pyrolysis and nitration. Industrial processing and the economics, design, and use of the equipment are based on these units' processes rather than consideration of each reaction separately.

Univalent (monovalent) Having a valency of one.

Universal indicator (multiple-range indicator) A mixture of indicator dyestuffs that shown a gradual change in colour over wide pH range. A typical formulation contains methyl orange, methyl red, bromothymol blue and phenolphthalein, and changes through a red, orange, yellow, green, blue and violet sequence between pH3 and pH10. Several commercial preparations are available as both solutions and test papers.

Unsaturated compound An organic compound that contains at least one double or triple bond between two of its carbon atoms. The multiple bond is relatively weak; consequently unsaturated compounds readily undergo addition reactions to form single bonds. Compare saturated compound.

Unsaturated solution *See* saturated solution.

Unsaturated vapour *See* saturated vapour.

Upper quartile If n measurements are arranged in order of magnitude from the least to the greatest, the ¾ (n-1) is the upper quartile. The upper quartile divides the area of the histogram in the ratio 3:1.

Upward capillary movement From a soil water table is another aspect of capillary movement. If the conducting capillaries are continuous uniform in size and unobstructed by the entrapped air and viscous colloidal substances, capillary upwards from a water table would be very effective.

Uranium A toxic radioactive silvery element of the actinoid series of metals. Its three naturally occurring radioisotopes, uranium-238 (99.283 per cent in abundance), uranium-235 (0.711 per cent), and uranium-234 (0.005 per cent), are found in numerous minerals including the uranium oxides pitchblende uraninite, and carnotite. The readily fissionable ^{235}U is a major nuclear fuel and nuclear explosive, while ^{235}U is a source of fissionable plutonium-239.

Uranium hexafluoride (UF_6) A crystalline volatile compound, used in separating uranium isotopes by differences in the rates of gas diffusion.

Uvula A large surface depression, upto about a kilometre across, that is found in limestone areas and formed by carbonation solution. It frequently results from the coalescence of several smaller depressions known as dolines. An uvula is intermediate between a doline and polje.

V

Vacuum A space containing gas below atmosphere pressure. A prefect vacuum contains no matter at all, but for practical purpose pascal, and hard (high) vacuum as below this. Ultra high vacuum is lower than 10^{-7} pascal.

Vacuum distillation Distillation of liquids under a reduced pressure so that the boiling point is lowered. Vacuum distillation is a common laboratory technique for purifying or separating unstable compounds *i.e.* ones that would decompose at their normal boiling point.

Vadose water Water in the ground that lies above the water table of level of total saturation of the rocks. Vadose water percolates down towards the water table through available joints or fissures in permeable rocks such as limestone. Caves eroded by Vadose water show many features typical of open channels on the surface, *e.g.* erosion kicks in the wall marking common water levels and meanders.

Valence (Valency) The combining power of any element or radical, equal to the number of hydrogen atoms that will combine with or displace one atom of the element. For simple covalent molecules the valence is obtained directly. C in CH_4 is tetravalent, N in NH_3 is trivalent. For ions the valence is regarded as equivalent to the magnitude of the charge, *e.g.* CA^{2+} is divalent, CO_3^{2-} is a divalent radical. The noble gases are zero-valent because they do not form compounds under normal conditions. As the valence for many elements is constant, the valence of some elements can be deduced without reference to compounds formed with hydrogen.

Valence electron An outer electron in an atom that can participate in forming chemical bonds.

Valeric acid *See* pentanoic acid.

Valine *See* amino acids.

Vaporization A gas formed by the vaporization of a solid or liquid. Some particles near the surface of a liquid acquire sufficient energy in collisions with other particles to escape from the liquid and enter the vapour, some particles in the vapour lose energy in collisions and re-enter the liquid. At a given temperatures, an equilibrium is established, which determines the vapour pressure of the liquid at that temperature.

Vapour density The ratio of the mass of a certain volume of a vapour to the mass of an equal volume of hydrogen (measured at the same temperature and pressure). Determination of vapour densities is one method of finding the relative molecular mass of a compound (equal to twice the vapour density).

Vapour pressure The saturated vapour pressure is the pressure of a vapour in equilibrium with its liquid or solid. It depends on the nature of solid of liquid and the temperature. In meteorology, the part of the total atmospheric pressure exerted by the water vapour present in the atmosphere. This increases as the water vapour increases and reaches a maximum when saturation is reached-the saturation vapour pressure.

Vapour movement of soil water Water vaporization relates to soils may be distinguished for convenience of discussion as internal and external.

Vapour pressure gradient This gradient is the difference in vapour pressure of two points a unit distance a part. Greater the difference, more rapid is the diffusion tends to become and the greater is the transfer of vapour water during a unit period.

Vapour pressure of soil water The water maintained in the air by the interchanges exerts a pressure called vapour pressure or aqueous tension.

Variable velocity When a body does not cover equal distance in equal intervals of time, however, small, the same direction in a straight line, its velocity is said to be variable.

Variation in loess soils In spite of physical and chemical homogeneity, of the loess as originally deposited, sometimes gives rise to soils of considerable diversity because of the climatic differences existent in various parts of 1 oessial area.

Variation in volume weight of soils Clay, clay-loam and silt loam surface soils may range from 1.00 to as high as 1.60 in volume weight depending on their conditions, while a variation from 1.20 to 1.80 may be found in sands and sandy loams.

Varve An annual layer of sediment laid down at the bottom of a lake in periglacial regions. Varve deposits have a banded appearance caused by alternating layers of different coloured liner and coarser sediments. Each have consists of a lighter band of coarser clay and slit sediment deposited in spring and summer after the thaw in cold countries.

Velocity It is the distance covered by the body in unit time in a particular direction.

Velocity ratio It is the ratio between the distance (D) through which the power acts and the distance (d) through which the weight is lifted. Thus V.R = D/d.

Vent An opening or conduit in the surface of earth through which volcanic material is ejected. A series of vents may form along a major fracture creating a fissure volcano. Volcanoes also develop with central vents and subsidiary vents on the aides of the cone.

Ventifcat A stone sculpted by the wind. Wind-blown sand flattens and smooth the surfaces as in mid-latitude or tropical deserts or in regions marginal to ice sheets.

Vermiculite A very light mica-based substance used as a support material for plant growth in some experiments where soil would be unsuitable. It can, for instance, be sterilised and mineral solutions of known composition be fed to the plants.

Vermilion *See* mercury (II) sulphide.

Vernier callipers It was invented by P. Vernier of Brussels and therefore it called Vernier Callipers. It consists of an auxiliary scale, called Vernier Scale which slides on the main scale. With its help, the fractional part of the main scale can be accurately measured. The vernier scale is divided such that its n divisions are equal to (n-1) or (n+1) divisions of the main scale.

Vinyl group The group CH_2: CH^-.

Volatile Easily converted into vapour.

Volcanic rock *See* extrusive rock.

Volcanism The processes responsible for upward movement of molten rock or magma through the earth's crust. It is responsible both for the volcanic eruptions and for the intrusion of the igneous rock within the crystal rock.

Volcano The point at which igneous material erupts through the earth's crust. Volcanoes usually consist of a conical hill or mountain formed from the molten rock or lava, ashes and rock fragments ejected through the volcanic vent. The exact shape of a volcano depends on many factors such as the viscosity of the lava and the amount of ash ejected. Volcanoes are described as active, dormant or extinct depending on the length of time since the last eruption. Extinct volcanoes can take many forms depending on the extent of subsequent erosions.

Volume composition of soils A silty loam surface is 45 per cent mineral, 30 per cent water, 20 per cent air and 5 per cent organic matter.

Volume and specified gravity of soils The specific gravity of soil represents 100 per cent soil material or the weight of absolutely solid soil. Volume indicates in comparison thereto the soil material actually present. The ratio of the volume weight to the specific gravity when multiplied by 100 becomes the percentage of the soil volume occupied by the soil particles.

Volume weight of soils In field, the volume weight of soils may be estimated by driving a cylinder of known volume into the ground and obtaining thereby a core of natural soil and then determining the amount of water that it contains. The amount of absolutely dried soil may be ascertained, dividing this by the weight of an equal volume of water gives the figure for volume weight.

Volt Symbol: V SI unit of electrical potential differences and e.m.f. defined as the potential difference between two points in a circuit between which a constant current of one ampere flows when the power dissipated in one watt 1V = II C^{-1}.

Voltmeter (cullometer) A device for determining electric charge or electric current using electrolysis. The mass m of material released is measured and this can be used to calculate the charge (Q) and the current (I) from the electro-chemical equivalent of the element using Q = m/z or I = m/zt.

Volumetric analysis One of the wet methods of quantitative analysis. It involves measuring the volume of a solution of accurately known concentration that is required to react with a solution of the substance being determined. The solution of known concentration (standard solution) is added in small portions in a burette. The process is called as titration and the equivalence point is the end point.

Vulcanizataion A process of improving the quality of rubber (hardness and resistance to temperature charges) by heating it with sulphur (at about 150°C). Accelerators are used to speed up the reaction. Certain sulphur compounds can also be used for vulcanization.

W

Washing soda *See* sodium carbonate.

Water (H_2O) A colourless liquid that freezes at 0°C and atmospheric pressure, boils at 100°C. In the gaseous state, water consists of single H_2O molecules. The atoms do not lie in a straight line the angle[2] between each hydrogen atom and the oxygen atom is 0.099 mm. It functions as a universal-weathering agent especially as water of crystallization.

Waste mantle It is a layer of unconsolidated weathered material lying above bedrock and below the ground surface. The parent material from which soil is formed and the soil itself is included in it. Also called Regolith.

Water absorbing capacity of colloidal clays Colloidal clays have a high water absorptive capacity. On drying, the colloids lose some of their absorbed water and cause shrinkage of soil. Colloidal particles also exhibit cohesive and adhesive properties. Soils with a high percentage of particles in colloidal state offer serious competition against plant roots for water and mineral nutrients in the soil solution.

Water and soil formation Torrential rain dislocates solid particles from rocks and puddles the surface of some materials. The debris is often transported a long distances by floodwaters of streams and rivers, and deposited as soil in low-lying areas, mounts of rivers and under the sea. Water frozen in rock crevices in cold regions breaks them open. Flowing water, waves on the sea and lake shores, and moving ice-glaciers in high mountain regions wear off rocks and cliffs, thus helping in soil formation.

Water erosion of soil Soil erosion caused by water can be distinguished in three forms, *viz.* (*i*) sheet erosion, (*ii*) rill erosion, and (*iii*) gully erosion.

Water erosion by rainwater Soil erosion caused by rainfall is the result of the application of energy from two distinct sources, namely (*i*) the falling rain drop, and (*ii*) the surface flow the energy of falling rain drop is applied slantingly or vertically from above while that of surface flow is applied more or less horizontally along the surface of the ground. The chief role of the falling raindrop is to detach soil particles, whereas that of surface flow (outside the rills and gullies) is to transport soil.

Water gap A narrow steep sided valley cut by a stream through a ridge of resistance rocks. It is commonly formed during the denudation of an area of domed sedimentary rocks where consequent streams maintain their course as they cut down through more resistant sedimentary strata.

Water gas A mixture of carbon monoxide and hydrogen produced when steam is passed over red-hot coke or made to combine with hydrocarbons, *e.g.* $C(s) + H_2O(g)\ CO(g) + H_2O + H_2(g)$

$$CH_4(g) + H_2O(g)\ CO(g) + 3H_2(g)$$

The production of water gas using methane is an important step in the preparation of hydrogen for ammonia synthesis.

Water holding capacity of soil The retentive capacity of a column of soil got water may be determined by fully saturating it and then allowing its d rain undisturbed by evaporation. This is also called field capacity and depends upon a number of conditions.

Water losses from soil Water lost by surface evaporation and transportation is replaced by the rise of capillary water from root zone. Finally a stage is reached when the water content becomes so low that the soil cannot supply water to the growing plant rapidly enough to maintain them turgid. Under such conditions, plants wilt permanently and the moisture condition of soil is termed as the permanent wilting percentage or 'wilting coefficient'.

Water potential A measure of energy for water in a plant. Water potential is fundamental thermo-dynamic function, such that water moves from a point of high water potential to one lower water potential.

Water power The energy that can be harnessed from running water. Early industrial sites used water power to turn water wheels and drive simple machinery *e.g.* the wool industry of East Anglina in the 14th century. Now this form of energy is harnessed as hydroelectric power to produce electricity that can be transmitted to factories a long way from the power stations.

Water softening The removal from water of dissolved calcium, magnesium and iron compounds thus reducing the hardness of water. The compounds are potentially damaging because they can accumulate in pipes and boilers and they react, with, and therefore waste soap. Temporary hardness can be removed by boiling the water. Permanent hardness can be removed by number of ways: by distillation, by addition of sodium carbonate, which causes dissolved calcium to precipitate out as calcium carbonate and by the use of ion-exchange products such as some substance called Permitit and colgon etc.

Water table The upper surface of the saturated rocks *i.e.* rocks in which all voids are filled with water. Water tables are rarely flat and tend to follow the relief of the land surface but with undulations smoothed out. They are generally deeper on the crest of a hill than in a valley bottom. The water table infects the ground surface water may emerge above ground in the form of springs, marshes or lakes. The upper surface of water accumulating in lower depths of saturating the underlying sand or gravel is known as 'water table'. This accumulated water furnishes supplies for shallow wells and springs.

Watt Symbol: W The SI unit of power defined as a power of one joule per second: $1 \text{ W} = 1\text{Js}^{-1}$.

Wax One of the group of water insoluble substance with a very high molecular weight. Waxed from protective coverings to leaves, stems, fruits, seeds, animal fur and the cuticles of insects, servings principally as water proofing.

Weak acid (or base) An acid or base that is not fully dissociated in solution.

Weathering The processes that result in the breakdown of rocks and minerals from either physical (mechanical) processes, such as freeze-thaw cycle heating and cooling; or chemical processes, such as carbonation, hydration, hydrolysis, solutions, organic weathering which consists of both mechanical and chemical weathering is caused by plants and animals, particularly when they borrow or release or leave acidic substance on the rocks. Lichens cause breakdown as they extract nutrients directly from the rocks.

Weathering front The basal surface of weathering.

Weathering process of soil Study of weathering and formation of the soil material line results not only interesting in itself, but is also necessary introduction to soil formation and classification.

Weber Symbol: Wb The SI unit of magnetic flux equal to the magnetic flux that linking a circuit of one turn produces an e.m.f. of one volt when reduced to zero at a uniform rate in one second: 1 wb = 1 V.

Weight of organic soil Weight of organic soil is lighter than the mineral soil. The volume weight compared with the mineral surface soils is surprisingly low, 0.20 to 0.30 perhaps being a fair estimate, a cultivated surface soil will usually fall within the limits of 1.25 to 1.45.

Weight of soil The absolute weight of soils varies from 50 to 120 lb. per cubic foot as compared with 65.5 lb. for one cubic foot of rich garden soil weighs about 70 lb.; of ordinary arable land 80 to 90 lb., and of dry sand 110 lb. The actual weight of soil may be expressed in terms of an acre-foot, referring to a volume of soil 1 acre in extent and one foot deep the weight of an acre-foot of surface mineral may vary from 30,00,000 to 40,00,000 pounds of dry substance.

Wentworth scale A method of classifying sedimentary rocks by grain size that was formulated by a U.S. scientist, C.K. Wentworth in 1922 and is the common scale used by the geologist. The scale ranges in geometric factors of 2 from particles of clay less than 1/256 mm in diameter in boulders of over 256 mm in diameter.

Weston cadmium cell A standard cell that produces a constant e.m.f. of 1.086 volts at 20°C.

Wet cell A type of cell, such as a car battery, in which the electroplate is a liquid solution. *See* also Leclanche cell.

White alkali *See* solanchank.

White lead *See* lead (II) carbonate hydroxide.

Wind It is an important carrying agent and when loaded with fine debris, exerts an advancing action in soil formation.

Wind erosion Wind erosion takes place normally in dry areas devoid of vegetation where the wind velocity is high. The soil particles on the land surface are lifted up and blown off in dust storms. When the velocity of the dust-bearing winds is retarded, coarser soils particles are deposited in dunes and the once fertile lands are rendered unfit for cultivation. In other places, fertile soil is drifted away by winds and the sub-soil is exposed, as a result of which the productive capacity of the soil is considerably reduced.

Wind gap A valley that cuts across a ridge and marks the route of a former river.

Wind velocities Wind velocities in India are on the whole low during the greater part of the year; in northern India, they are low practically throughout the year. Even in southern India, except in a few favourable localities along the coast, the wind velocity is high during monsoon months only when very little irrigation is required.

Winds and soil formation Sand storms in desert, semi-arid regions and dry river beds, and high winds on the sea shore have both erosive and transporting action. Moving wind also has a drying effect, especially in region of low average humidity.

Witherite A mineral form of barium carbonate $BaCO_3$.

Wohler's synthesis A synthesis of urea from inorganic compounds. Urea was produced by evaporating NH_4NCO (ammonium cyanate).

Wolfram Previous name for tungsten. Wood alcohol–Seem methanol Wood's metal–A fusible alloy containing 50 per cent bismuth,

25 per cent lead and 12.5 per cent tin and cadmium. Its low melting point (70°C) makes its use possible in fire protection devices.

Work When a force acts upon a body and the point of application of force moves, work is said to be done by the force.

Work done against friction There are two situations for this work– (1) in a horizontal plane, and (2) rough inclined plane. In the rough inclined, either the body is dragged or the body is prevented from moving down.

X

Xenolith A fragment of rock that is contained within an igneous instruction from which it differs in origin, structure, composition etc. The rock may be a fragment of the rock through which the magma of the igneous intrusion forces its way.

Xenon A colourless, odourless monatomic element of the rare gas groups, now known to form compounds. It occurs in trace amounts in air. Symbol: Xe.

X-radiation An energetic form of electro-magnetic radiation. The wavelength range is 10^{-1} to 10^{-8} m. X-rays are produced by absorbing high energy electrons in matter. The radiation can pass through matter to some extent (hence its use in medicine and industry for investigating internal structures).

X-ray crystallography The study of the internal structure of crystal, using the technique of X-ray diffraction.

X-ray diffraction A technique to determine crystal structure by directing X-rays at the crystals and examining the diffraction pattern obtained.

Y

Young soils The solid in flooded areas is subject to frequent deposit of fresh alluvial material. It always remains youthful in terms of years and stage of development. Young soils in flooded areas are usually very fertile, whereas those along stony hill slopes may be of little agricultural valve.

Z

Zeolite A member of group of hydrated aluminosilicate minerals, which occur in nature and are also manufactured for their ion-exchange properties. They are used for water softening and for sugar refining. The zeolites have an open crystal structure and can be used as molecular sieves.

Zero error When the jaws of vernier are closed, the zero of the vernier scale should coincide with the zero of the main scale. Otherwise, this error is called zero error. The zero error is either +ve or –ve.

Zero order It describes a chemical reaction in which the rate of reaction is independent of the concentration of a reactant:

$$\text{rate} = k\,(X)^{o}$$

The concentration of the reactant remain constant for a period of time, although other reactants are being consumed.

Zero point energy The energy possessed by atom and molecules of a substance at absolute zero (OK).

Zeuge A desert landform consisting of an upstanding regular rock undercut around its edges by erosion by wind blown sand.

Ziegler process A method for the manufacture of high-density polythene using catalyst of titanium (IV) chloride and triethyl aluminium (A1$(C_2H_5)_3$ under slight pressure. The mechanisms involved formation of titanium alkyls–$TiCl_3$ (C_2H_5) which co-ordinate the orbitals of titanium with the bond of ethane. The chain length and hence the density of the polymer can be controlled.

Zinc 420°C, b.p. 908°C, r.d. 7.8, p.n. 30, r.a., 65.38 A bluish-white brittle metal. At a temperature between 100°C and 150°C, it can be rolled into sheets and drawn out into wires.

Zincate A salt containing the ion ZnO_2.

Zinc-blend structure A form of crystal structure that consist of a zinc atom surrounded by four sulphur atoms arranged tetrahedrally, each sulphur is similarly surrounded by four atoms of zinc.

Zinc oxide (ZnO) A compound formed by the thermal decomposition of zinc nitrate or carbonate; it is white powder when cold, yellow when hot.

Zinc sulphate ($ZnSO_4$) A white crystalline solid prepared by the action of dilute sulphuric acid on either zinc oxide or zinc carbonate.

Zircon *See* zirconium.

Zirconium A transition element that occurs in a gem-score ($ZnSiO_4$). It is used in some strong alloy steels. Symbol: Zr; m.p. 1850°C, b.p. 4380xc; r.d. 6.5; p.n. 40; r.a.m. 91.22.

Zonal soil A soil that is well-developed and owes its character, especially profile development, mainly to the influence of climatic and/or vegetation. Other soil forming profile are less important. The term applies only to mature free draining soils *e.g.* podzols, chemozems, which develop on parent material that is neither acid nor very alkaline.

Zonal soil groups With these soils, erosion is not a serious factor. Their climatic influences are uniform and continuous. Such soil groups are spoken as zonal. They constitute the great soil groups of the world. It is a soil that is well-developed and has its own character, especially profile development, mainly to the influence of climate and/or vegetation. Other soil forming factors are less important. The term applies to mature free draining soils (*e.g.* podzols, chernozems) developed on material that is neither very acid nor very alkaline.

Zwitterion An ion with a positive and a negative charge. Amino-acids can form zwitterions; the amino groups has the form NH_3 and the acid group is ionised to –COO. It is also called ampnolyte ion and is an ion that has both a positive and negative charge on the same species.